Xiaoyu Zhang

Crosswind stability of vehicles under nonstationary wind excitation

Karlsruher Institut für Technologie
Schriftenreihe des Instituts für Technische Mechanik

Band 28

Eine Übersicht aller bisher in dieser Schriftenreihe erschienenen Bände finden Sie am Ende des Buchs.

Crosswind stability of vehicles under nonstationary wind excitation

by
Xiaoyu Zhang

KIT Scientific Publishing

Dissertation, Karlsruher Institut für Technologie (KIT)
Fakultät für Maschinenbau
Tag der mündlichen Prüfung: 02. April 2015

Impressum

KIT Scientific Publishing

Karlsruher Institut für Technologie (KIT)
KIT Scientific Publishing
Straße am Forum 2
D-76131 Karlsruhe

KIT Scientific Publishing is a registered trademark of Karlsruhe Institute of Technology. Reprint using the book cover is not allowed.

www.ksp.kit.edu

Print on Demand 2015

ISSN 1614-3914
ISBN 978-3-7315-0376-7
DOI 10.5445/KSP/1000047001

Crosswind stability of vehicles under nonstationary wind excitation

---

Zur Erlangung des akademischen Grades

Doktor der Ingenieurwissenschaften

der

Fakultät für Maschinenbau
Karlsruher Institut für Technologie (KIT)

genehmigte
Dissertation

von

M. Eng. Xiaoyu Zhang
aus Jiangsu, China

Tag der mündlichen Prüfung:	02. April 2015
Hauptreferent:	Prof. Dr.-Ing. Carsten Proppe
Korreferent:	Prof. Dr.-Ing. habil. Michael Hanss

Vorwort

Die vorliegende Dissertation entstand während meiner Tätigkeit als Stipendiat in den Jahren 2010 bis 2014 am Institut für Technische Mechanik des Karlsruher Instituts für Technologie.

Mein ganz besonderer Dank gilt Herrn Prof. Dr.-Ing. Carsten Proppe für die Anregung zu dieser Arbeit, deren Betreuung und die Übernahme des Hauptreferates. Seine großzügige Unterstützung und Förderung habe ich stets sehr geschätzt.

Ebenso bedanke ich mich sehr herzlich bei Herrn Prof. Dr.-Ing. habil. Michael Hanss vom Institut für Technische und Numerische Mechanik der Universität Stuttgart für die Übernahme des Korreferates und die wertvollen Anmerkungen.

Bei Herrn Prof. Dr.-Ing. Peter Gratzfeld bedanke ich mich für die Übernahme des Prüfungsvorsitzes, für das entgegengebrachte Interesse an meiner Arbeit sowie für die Durchführung des Promotionsprozesses.

Ebenfalls danke ich Herrn Prof. Dr.-Ing. Wolfgang Seemann, Herrn Prof. Dr.-Ing. habil. Alexander Fidlin, Herrn Prof. Dr.-Ing. Dr. h.c. Jens Wittenburg und Herrn Prof. Dr.-Ing. Walter Wedig für das Interesse an meiner Arbeit, die wissenschaftliche Begleitung und ihre Diskussionsbereitschaft.

Allen Kollegen am Institut für Technische Mechanik möchte ich herzlich für die angenehme Arbeitsatmosphäre danken. Ganz besonders danke ich Herrn Dr.-Ing. Rugerri Toni Liong, Dr.-Ing. Dipl.-Math. Alexander Karmazin, Dipl.-Ing. Maximilian Geißendörfer, Frau M.-Eng. Han Hu, Dr.-Ing. Nicole Gaus und Dr.-Ing. Heike Vogt für die große Hilfe und interessanten Diskussionen während unserer gemeinsamen Zeit am ITM, es war eine tolle Zeit. Herrn Dr.-Ing. Rugerri Toni Liong und Dipl.-Ing. Maximilian Geißendörfer danke ich nochmals für das Korrekturlesen meiner Dissertation.

Abschließend möchte ich mich ganz herzlich bei meinen Eltern, meiner Schwester und meiner lieben Ehefrau Zhiyi bedanken. Ohne ihre unendliche Unterstützung, ihre Geduld und ihre Rücksichtnahme wäre diese Arbeit so nicht möglich gewesen. Vielen Dank für alles!

Karlsruhe, den 08. April 2015
Xiaoyu Zhang

Kurzfassung

Die ausreichende Seitenwindstabilität ist ein wichtiges Kriterium bei der Zulassung von Fahrzeugen. In den vergangenen Jahren ereigneten sich viele windbedingte Unfälle von Schienen- und Straßenfahrzeugen. Obwohl viele Bemühungen auf die Untersuchung der Seitenwindstabilität von Fahrzeugen aufgewendet worden sind, ist die Erforschung von instationärer Seitenwindstabilität in einem realistischeren Windszenario eine große Herausforderung. In der vorliegenden Arbeit wird eine probabilistische Methode verwendet, um die Seitenwindstabilität von Fahrzeugen zu analysieren. Weiterhin sind die instationären Aspekte für die Risikobewertung von Fahrzeugen unter starken Seitenwind untersucht worden.

Aufgrund der Existenz der Turbulenz ist der natürliche Wind immer ein stochastischer Prozess. Realistische Annahmen über die Natur des Seitenwinds müssen dessen instationären Charakter berücksichtigen. Basierend auf einer Analyse von natürlichem Wind wurde ein stochastisches Böenmodell mit instationärer Windturbulenz für die Risikoanalyse der Seitenwindstabilität von Fahrzeugen vorgeschlagen und verwendet. Gemessene Winddaten wurden verwendet, um das instationäre Böenmodell zu kalibrieren.

Außer dem instationären Windmodell wurde der instationäre Fahrzeugzustand in unterschiedlichen Szenarien in Betracht gezogen. Schienenfahrzeuge, die in Kurven und auf Geraden mit unterschiedlicher Fahrzeuggeschwindigkeit unter der instationären Windanregung fahren, werden untersucht. Straßenfahrzeuge werden in verschiedene Kategorien (z.B. kleine Personenkraftwagen, Lastkraftwagen, Busse, Kleintransporter und Personenkraftwagen mit Anhänger, usw.) eingeteilt. Für jede Fahrzeugklasse wird ein entsprechendes repräsentatives Fahrzeugmodell mit dessen nichtlinearen aerodynamischen Koeffizienten identifiziert. Das Fahrermodell sowie die Straßenbedingungen (trocken/nass) werden ebenfalls betrachtet.

Die instationäre Windanregung für ein räumlich ausgedehntes fahrendes Fahrzeug wird durch Verwendung des äquivalenten Windspektrums, welches die aerodynamische Admittanz-Funktion berücksichtigt, berechnet. Darüber hinaus wird die instationäre Windanregung durch die Hilbert-Huang Transformation (HHT) analysiert. Diese ermöglicht die Zerlegung der Windkraft in mehrere empirische Moden und zeigt die wichtigen Eigenschaften der instationären Anregung durch instantane Frequenzen auf.

Da die Windanregung auf dem Fahrzeug ein stochastischer Prozess ist, muss eine Risikoanalyse durchgeführt und die Versagenswahrscheinlichkeiten berechnet werden. Die Amplitude und Dauer der Böe sowie die nichtlineare aerodynamischen Koeffizienten werden als

Zufallsvariablen betrachtet. Die Berechnung der Versagenswahrscheinlichkeiten kann entweder auf analytischen Methoden (z.B. FORM, SORM) oder auf Monte Carlo Simulation mit der Varianzminderung (z.B. LS, SS) basieren.

Basierend auf dem vorgeschlagenen Modell und der verwendeten Methoden kann der Einfluss des Winds und der Fahrzeuggeschwindigkeit sowie weiterer Parameter (z.B. Windturbulenz, Kurvenneigung, Straßenbedingung) auf die Seitenwindstabilität von Fahrzeugen identifiziert werden. Gegenmaßnahmen wie Windzäune und Geschwindigkeitsbeschränkungen können auf einer objektiven Basis beurteilt werden. Mit Hilfe der Simulationsergebnissen kann ein Windwarnsystem für Fahrzeuge abgeleitet werden, das auf Risikoniveaus basiert.

Abstract

Sufficient crosswind stability has become an important criterion in the approval process of ground vehicles. During the past years, many wind-induced accidents of railway and road vehicles have been reported. Although a lot of efforts have been spent on investigating the crosswind stability of vehicles, the research of nonstationary crosswind stability is a great challenge. In the present work, a probabilistic method is used to analyze the crosswind stability of vehicles and the nonstationary aspects for risk assessment of vehicles under strong crosswinds have been studied.

Due to the existence of turbulence, natural wind is always a stochastic process. Realistic assumptions on the nature of crosswind have to take its nonstationary character into account. Based on the characteristic analysis of natural wind, a stochastic gust model with nonstationary wind turbulence has been proposed and utilized for the risk analysis of crosswind stability of vehicles. Wind data that has been measured at the exposed site are used to calibrate the nonstationary gust model.

The resulting nonstationary wind excitation for a moving vehicle with spatial extension has been calculated by using an equivalent wind spectrum, which takes the aerodynamic admittance function into account. To analyze the nonstationary wind excitation, Hilbert-Huang transform (HHT) has been applied. It allows for a decomposition of the wind force into several intrinsic mode functions (IMFs) and shows important characteristics of the nonstationary excitation by instantaneous frequencies.

Besides the nonstationary wind model, the nonstationary vehicle state in a variety of different physical scenarios has also been taken into consideration. Railway vehicles running on curved and straight track with varying vehicle speed under nonstationary wind excitation are studied. Road vehicles are classified into different categories: small passenger cars, trucks, buses, vans and car-trailers. For each vehicle class, a corresponding worst-case vehicle model together with its nonlinear aerodynamic coefficients is identified. The driver model as well as the road conditions (dry/wet) are also considered.

As the wind excitation on the vehicle is a stochastic process, a risk analysis has to be carried out and failure probabilities have to be computed. The gust amplitude and duration as well as the nonstationary aerodynamic coefficients are considered as random variables. Computation of the failure probabilities can be either based on the analytical methods (e.g. FORM, SORM) or Monte Carlo methods with variance reduction (e.g. line sampling (LS), subset sampling (SS)).

Based on the proposed model and approach, the influence of wind and vehicle speed as well as other parameters (e.g. wind turbulence, track cant, road condition) on the crosswind stability of vehicles can be well identified. Countermeasures such as wind fences and vehicle speed limitations can be judged on an objective basis. For road vehicles, the simulation results allow to determine the parameters (e.g. critical wind speeds) of a wind warning system for traffic based on a risk level.

Contents

List of symbols

Indizes

dyn	dynamic
lf	left-front
lr	left-rear
rf	right-front
rr	right-rear
R	Rayleigh distribution
sta	static
t	trailer
W	Weibull distribution
w	wind

Symbole

α	slip angle of the wheel
α_ω	wind angle
$\bar{F}$	mean wind force
$\bar{T}$	mean gust duration
$\bar{\beta}_\omega$	mean part of the relative wind angle
β	scale parameter
β'_ω	fluctuating part of the relative wind angle
$\beta_\omega(t)$	total relative wind angle

δ_i	wheel steering angle
$\delta(t)$	delta function
δ_f	steering angle of the front wheel
δ_Q	safety margin
δ_r	steering angle of the rear wheel
$\dot{\tilde{y}}_l$	train velocity in the wind coordinate system, lateral direction
$\dot{\tilde{y}}_v$	train velocity in the wind coordinate system, vertical direction
$\hat{t}$	instant of time
λ^+	a failure rate
λ_0	wheel conicity
$\hat{\mathbf{z}}$	grouped random variables
$\hat{\mathbf{z}}_{MPP}$	most probable point (MPP)
$\mathbf{e}_\alpha$	unit vector of the important direction
$\mathrm{Coh}(f)$	coherence function
$\omega(t)$	instantaneous frequency
ω	wheel angular speed
$\overline{V}_{res}$	mean relative wind speed
ϕ	roll angle of the vehicle
ρ_L	air density
τ	time lag
$\theta(t)$	instantaneous phase angle
θ_t	cant angle
θ	pitch angle of the vehicle
θ_e	hitch angle

$\tilde{A}$	normalized gust amplitude
$\tilde{T}$	normalized gust duration
$\tilde{x}(t)$	Hilbert transform of $x(t)$
ε_t	a white Gaussian noise process
A	gust amplitude
$A(t)$	instantaneous amplitude of $x(t)$
$a(z,t)$	modulating function
a, b, ζ	scale, location and shape parameters
A_0	threshold value of the gust amplitude
A_C	characteristic area of the vehicle
$A_i(t), B(t)$	time dependent coefficients for the NONAR model
a_x	longitudinal acceleration
a_y	lateral acceleration
a_z	vertical acceleration
$a_{y,p}$	permissible lateral acceleration
C_F	dimensionless aerodynamic coefficient
$c_i(t)$	component of each IMF
C_s	damping coefficient of the suspension
C_s	longitudinal cornering stiffness
C_α	lateral cornering stiffness
$e(t)$	step function
e_1	distance between the hitch point and rear axle of the car
e_2	distance from the hitch point to front axle of the trailer
f	frequency

f	probability density function
F'	fluctuating wind force
F'_h	reaction forces of F_h
$F(t)$	wind force
$f(u)$	probability density function
F_0	mean wind force (static vehicle)
f_0	cut-off frequency of the road surface
f_c	upper cut-off frequency
$F_L(t)$	lifting wind force
f_r	rolling resistance coefficient
F_s	suspension force
$F_S(t)$	side wind force
F_u	cumulative distribution function
F_x	Tire force in x direction
F_y	Tire force in y direction
F_z	normal tire force
$f_{\tilde{A}}$	probability density function of $\tilde{A}$
F_{hxt}	hitch force acting on the trailer in x direction
F_{hyt}	hitch force acting on the trailer in y direction
F_h	hitch force acting on the car
f_{st}	Strouhal frequency
F_w	wind force acting on the vehicle
F_{xw}	longitudinal tire force in the wheel plane
F_{yw}	lateral tire force in the wheel plane

$g(\hat{\mathbf{z}})$	performance function
$G(x)$	Pareto distribution
$G(x)$	distribution of the maximal values
$h(\omega)$	marginal spectrum
$H(\omega, t)$	Huang spectrum
$h_F(\tau)$	aerodynamic weighting function
H_T	height of the vehicle
h_t	track cant
$h_{e,p}$	permissible cant deficiency
h_{eq}	equilibrium cant
h_e	cant deficiency
$h_{l,u}$	aerodynamic weighting function for lift force in the lateral direction
$h_{l,w}$	aerodynamic weighting function for lift force in the vertical direction
$h_{s,u}$	aerodynamic weighting function for side force in the lateral direction
$h_{s,w}$	aerodynamic weighting function for side force in the vertical direction
I_F	a indicator function
I_w	moment of inertia for the wheel
I_{xx}	Rolling moment inertia
I_{yy}	Pitching moment inertia
I_{zz}	Yawing moment inertia
K_s	spring stiffness of the suspension
K_t	normal stiffness coefficient of the tire
L	characteristic length of the cross-section
L_C	characteristic length of the vehicle

l_f	distance from the front axle to the gravity center
l_r	distance from the rear axle to the gravity center
L_T	length of the vehicle
l_T	nominal distance between the wheel-rail contact points of a wheelset
l_t	track distance measured on the top track plane
m	vehicle mass
$M(t)$	wind moment
m_t	mass of the trailer
M_w	Wind moment acting on the vehicle
N	number of IMFs
N	occurrence frequency of gust amplitudes
N	simulation times
N	total number of the wind data
n	total number of wind speed intervals
P	Cauchy principal value
$P_f, P(F)$	failure probability of the vehicle
P_G	failure probability for a single gust event
P_{f,T_f}	failure probability for an independent time interval
Q	wheel load
q_1	intermediate variable
q_2	intermediate variable
Q_r	sign for the change of wheel force
R	auto-correlation function
R	radius of the curved track

r	yaw angle of the vehicle
$r(t)$	residue of the signal
r_0	nominal rolling radius
r_t	yaw angle of the trailer
R_w	wheel radius
$R_{\varepsilon X}$	cross-correlation function for X and ε_t
S_i	longitudinal wheel slip ratio
S_t	Strouhal number
S_z	road roughness coefficient
$S_{F'F'}(f)$	power spectral density of the wind force
$S_{u'u'}(f)$	wind power spectrum
T	gust duration
T	track width
T_b	braking torque of the wheel
t_B	time at the maximal gust speed
T_d	driving torque of the wheel
u	threshold value
$u'_u(t), u'(t)$	wind turbulence in the mean wind direction
$u'_v(t)$	wind turbulence in the lateral direction
$u'_w(t)$	wind turbulence in the vertical direction
$U(t)$	total wind speed in space
u^*	friction velocity
u_0	mean wind speed
$u_0(z)$	mean wind speed at height z

$u_B(t)$	coherence function
$u_u(t)$	longitudinal wind speed
$u_v(t)$	lateral wind speed
$u_w(t)$	vertical wind speed
u_{wx}	longitudinal wheel velocity in the wheel plane
u_{wy}	lateral wheel velocity in the wheel plane
v_0	mean vehicle speed
v_p	permissible vehicle speed
V_x	vehicle velocity along x direction
V_y	vehicle velocity along y direction
$V_{res}(t)$	relative wind speed
V_{xt}	longitudinal velocity of the trailer
V_{yt}	lateral velocity of the trailer
$w(a,b)$	continuous wavelet transform (CWT) of a real signal
$w_{j,k}$	discrete wavelet transform (DWT) of a signal
X	nonstationary time series
x	stationary time series
$x(t)$	original nonstationary signal
y_0	lateral displacement of the road vehicle
z	height in the space
$z(t)$	analytical signal
$z(t)$	road irregularity for the wheel
z_0	surface roughness length
Z_s	displacement of the sprung mass

Z_u	displacement of the unsprung mass
$\chi_F^2(f)$	aerodynamic admittance
$\Gamma(t)$	gamma function
κ_r	reduction factor of the road adhesion
$\mathbb{Z}(z,f)$	an orthogonal increment process
μ_i	friction coefficient of the road
μ_R	mean value of the Rayleigh distribution
μ_W	mean value of the Weibull distribution
$\psi(t)$	mother wavelet
$\Psi(z,t)$	a stationary Gaussian stochastic process
$\psi_{j,k}(t)$	discrete wavelet function
σ_u	standard deviation of the wind speed
$\sigma_{Ln(\tilde{T})}$	standard deviation of the log-normal probability function
σ_R	variance of the Rayleigh distribution
σ_W	variance of the Weibull distribution
$\Theta(z,f)$	power spectral density of $\Psi(z,t)$
$\Theta(z_1,z_2,f)$	cross-spectral density function
φ_i	independent random phase angle
φ	phase angle

List of Abbreviations

AC	Acceleration
AR	Autoregressive model
ARMA	Autoregressive moving average model
DE	Deceleration
EPSD	Evolutionary power spectral density
FORM	First Order Reliability Method
FT	Fourier transform
HHT	Hilbert-Huang Transform
IMF	Intrinsic mode function
LS	Line sampling
MA	Moving-average model
MAX	maximum method
MPP	Most Probable Point
NONAR	Nonstationary-Autoregressive
PDF	probability density functions
POT	peaks over threshold method
PSD	Power spectral density
Turb	Turbulence

1 Introduction

In recent years, crosswind stability has attracted more and more attention. It has become an important criterion in the approval process of railway and road vehicles. Strong crosswind may lead to accidents or at least to a discomfort for the driver of road vehicles. In the past, more than 30 wind-induced accidents of railway vehicles were reported around the world (e.g. Japan, Belgium, Italy, Switzerland and China) [161] [14], most of them happened on narrow gauge tracks at wind speeds that were considerably high [161]. Incidents for overturning of road vehicles or other accidents caused by strong crosswind are frequent and have also been reported [87] [45]. An example of wind-induced accidents of railway and road vehicles can be seen in Figure 1.1.

(a) railway vehicle [15]

(b) road vehicle [194]

Figure 1.1: Examples of the wind-induced accidents

Generally speaking, the effects of crosswinds on railway and road vehicles can be basically classified into two groups, that is, the effect due to high crosswind and the effect due to low crosswind, respectively. In high crosswinds, railway and road vehicles (especially high-sided lorries and large vans) are blown over on a regular basis [7] [15]. For lower wind speeds, although the effects of crosswinds are not so obvious, it can still lead to operational difficulties. Cooper [58] has shown that there is a possibility of the excitation of suspension modes for high-speed trains in relatively low wind speed conditions. Moreover, the drivers of road vehicles may experience significant fatigue when driving a long time in gusty winds.

In many countries, including Germany, methods for crosswind stability analysis are based on numerical simulation and sometimes on wind tunnel tests to verify the numerical results. However, due to the high cost of wind tunnel experiments and the nonstationary characteristic of natural crosswinds as well as a large diversity of vehicle types, the simulation of vehicles running in various kinds of realistic wind scenarios by experimental methods seems very difficult. Therefore, numerical simulations and calculations are becoming more and more popular. For numerical simulation, the quality of the safety evaluation depends on the accuracy of applied models. Vehicle dynamics can be simulated with high accuracy by commercial multi-body dynamics software like Adams, Simpack and etc. On the other hand, the aerodynamic excitation acting on the vehicle, which highly depends on vehicle shapes and wind scenarios, is usually a nonstationary process and can be evaluated usually with a low accuracy [33]. Besides, vehicle parameters (e.g. vehicle mass, side area and height of gravity center) and road/track conditions (e.g. dry or wet road, irregularities) are usually uncertain, making it quite difficult to obtain an accurate assessment of the crosswind stability.

1.1 Motivation of the work

In order to assess the crosswind stability of vehicles in a more realistic way, more work has to be carried out. On the one hand, current developments in vehicle engineering have been showing a trend of faster, more energy efficient and more comfortable vehicles with a high capacity of passenger transportation [204]. These trends require that the vehicle has a much better handling stability. However, the demands for light-weight constructions and higher driving velocities are in conflict with crosswind stability. On the other hand, accidents or injuries may occur as a result of strong crosswinds. For road vehicles, a secondary damage may also happen (e.g. an obstruction of the traffic or a following accident due to the blown-down truck). All these factors make it important to study the crosswind stability for ground vehicles, especially for high-speed railway vehicles and road vehicles that are easily influenced by wind (e.g. light trucks, caravans).

A significant challenge for studying the crosswind stability of vehicles is to carry out a nonstationary analysis (including the nonstationary crosswind, nonstationary vehicle state and nonstationary wind excitations). Traditional methods for the research of crosswind stability are based on deterministic parameters. The underlying assumption for these methods is that all the parameters such as gust amplitude and gust duration are assumed as deterministic values. In reality, this is not always the case. Due to the existence of turbulence, wind is always a stochastic process. Realistic assumptions on the nature of crosswinds have to take its nonstationary character into account. Moreover, owing to the strong nonlinear characteristic of the relationship between the vehicle dynamics and strong crosswinds, the assumed deterministic gust models seem hard to quantify and may lead to unreliable results. Therefore, a realistic wind model with nonstationary wind turbulence as well as the wind effects on the moving vehicle have to be investigated.

Besides the nonstationary wind model, more realistic vehicle states in various wind scenarios should also be taken into consideration. Irregularities or other geometrical characteristic of the track (e.g. straight line and curved line) for railway vehicles are needed to be considered. For road vehicles, different types of vehicles together with the driver models and road conditions are necessary to be distinguished.

All in all, this work is mainly motivated by the nonstationary aspects for crosswind stability of vehicles. In order to study the stability of vehicles under strong crosswinds in a more realistic way, a nonstationary wind model as well as nonstationary vehicle state (e.g. varying vehicle speed) have to be considered.

1.2 Literature review

1.2.1 Nonstationary wind

Crosswind stability for railway and road vehicles is usually calculated based on a stationary situation or at least on wind tunnel experiments that are mostly carried out with a static vehicle model. Nonstationary excitation due to wind turbulence occurs if the vehicle accelerates or decelerates. It has been realized that nonstationary wind has a great influence on the vehicle stability especially when the vehicle speed is high. Increasing vehicle speed relative to the wind speed will make the energy content of the spectrum move into a higher frequency range [59]. In addition, nonstationary wind excitation may come out due to a sudden gust. For instance, in a tunnel-exit or an overtaking scenario, a strong gust may occur just after the vehicle has left the tunnel or passed another vehicle.

Many of the foregoing research encompassing modeling and simulation of wind effects on structures have tacitly assumed that wind is a Gaussian and stationary process. This assumption is quite valid for wind forces that involve integral effects of a random pressure field over large areas [92]. Nonetheless, it has been noted that the structure under separated flows experience strong non-Gaussian effects in regions of negative pressure (leeward side) [155]. Non-Gaussian characteristics in wind speed fluctuations occurs at the height near ground. For low structures immersed in strong turbulence, they may experience non-Gaussian pressure fluctuations even on their windward side [92]. Similar effects have also been found by Holmes(1981) [99] and Kareem(1994) [118]. Moreover, nonstationary characteristics of the wind has been noted in extreme wind events (e.g. thunderstorm) [47]. The wind flow caused by extreme wind events varies significantly from that in the traditional atmospheric boundary layer in terms of its rapid time-varying mean wind speed and spatially strongly correlated wind fluctuations [83]. The time-varying amplitude and frequency components embedded in these processes make the stationary assumption inappropriate.

In consideration of the deficiency of stationarity assumption, methods on modeling nonstationary wind have been studied. Until now, different methods for the analysis of nonstationary wind turbulence have been proposed, such as the evolutionary power spectral density (EPSD) method, wavelet method and autoregressive-moving-average-model (ARMA) method [157] [199] [116] [122] and [147]. Among these methods, one of the most attractive is wavelet analysis, which has been successfully applied in many fields, including communication systems and image analysis. One of the most interesting features of the wavelet method is the fact that the wavelet coefficients deduced from time sequential data represent the energy in time and frequency domains, which is nearly impossible by the traditional Fourier transform. The traditional Fourier transform is based on the hypothesis of stationary process and cannot describe the local transient features due to its averaging over the whole duration of the signal [91] [112].

Another attractive method to produce nonstationary wind turbulence is the method based on recursive digital filtering with a ARMA model. It has been proved to be the most efficient method in terms of computational efforts and computer resource demands [149]. Rossi [170] assessed the performance of different numerical procedures for autoregressive model (AR), ARMA and gave a detailed description for ARMA modeling while Maeda and Makino [137] used ARMA to successfully model the turbulence of gusty wind. In this work, the nonstationary autoregressive (NONAR) method based on ARMA model is applied to produce the nonstationary wind turbulence, cf. Chapter 2.

In addition, to simulate extreme wind conditions, gust scenarios are also usually utilized. The nonstationary gust scenario is considered as a combination of two parts, i.e. the nonstationary wind turbulence and the stochastic gust. For the gust model, there are various different gust shapes which have been proposed in the literature so far, such as '1-cos' model [193] [127], 'Chinese hat' [84] and 'rugby-ball' [54]. They are obtained either by convention or by selecting a reference gust out of a statistical distribution. However, due to the strong nonlinear characteristic of the relationship between wind speed and vehicle dynamics, the simplification proposed by assuming a gust with determined amplitude and duration seems to be difficult to quantify [161]. A recently proposed method for simulating the gust that has been developed for reliability analysis of wind turbines is the so called constrained-simulation approach. The main advantage of this method is that the turbulent wind process with the superimposed gust characteristic is statistically indistinguishable from the natural turbulent wind process, which can be seen in [203] [26].

1.2.2 Railway vehicles under crosswind

Currently, vehicle dynamic responses caused by nonstationary wind as well as the nonstationary vehicle state have attracted a lot of attention, cf. [11] [12] [14] [162] and [195]. Based on Taylor's frozen turbulence hypothesis, Cooper (1984) [59] developed a random-process model for the turbulent wind with respect to a moving vehicle. The numerical results of wind power spectral density (PSD) for a moving vehicle has been firstly introduced. Later, in order to study the characteristic of unsteady aerodynamic forces, Baker (1991) [11] considered the wind-induced force in the frequency, amplitude and time domains, respectively. It has been pointed out by Baker that the quasi-steady assumption for the crosswind analysis is only valid under the condition that the gust duration is greater than about 5 s for lorries and trains, or greater than 1 s for cars. Additionally, it has been also indicated by Baker that the wind force and moment predicted by convolution integral analysis in the time domain are in qualitative agreement with experimental data.

Considering the real effect of wind fluctuations on the vehicle (e.g. attenuation or lag), a parameter named 'aerodynamic admittance', which is defined as a ratio between the force spectrum and wind spectrum in the frequency domain, is usually proposed. The corresponding name for aerodynamic admittance in the time domain is 'weighting function'. In order to determine the aerodynamic admittance for a railway vehicle, a number of experiments have been undertaken in the past, see [18] [188] and [15]. Baker [18] determined the unsteady forces on the train due to crosswinds through the concept of aerodynamic weighting function from experimental data of aerodynamic admittance. The aerodynamic excitation on high-speed vehicles produced by an unsteady crosswind were reported.

Recent results about nonstationary crosswind force to the moving vehicle can be found in [12] [14] and [196]. Baker [14] developed a method for the simulation of unsteady aerodynamic forces on a moving vehicle by the use of weighting functions. Characteristics of the fluctuating crosswind and the corresponding nonstationary force were described. Besides, Thomas [196] studied the unsteady performance of a high-speed vehicle running in strong crosswinds. A simulation of a railway vehicle under the influence of curved tracks has been carried out. These investigations offer a good reference for the study of nonstationary crosswind stability. However, they usually presented mean results based on stochastic wind models with stationary wind turbulence. The risk analysis based on a nonstationary wind scenario is not considered.

In [34], a study based on reliability analysis for crosswind stability of railway vehicles has been carried out. The probabilistic characteristic wind curves (PCWC) which shows the failure probability of vehicles running under strong crosswind conditions were presented. A similar investigation can be found in [202], where a probabilistic approach for assessing the crosswind stability has been proposed. These studies took the uncertain characteristics of crosswinds into account and calculated the corresponding failure probabilities for vehicles. However, they also have not taken the nonstationary character of the natural wind as well as the nonstationary vehicle state (e.g. acceleration or deceleration) into consideration.

In addition, vehicle ride instability may occur when the vehicle speed is high and in some cases it could also affect the vehicle ability to withstand crosswind [195]. As one of the most important parts of the railway system, track geometry has a significant influence on the vehicle dynamics. For straight track, the vehicle lateral dynamics can be affected by the track irregularities as well as the running behavior of the vehicle. When the vehicle runs on curved track, situations will become more complicated. A vehicle running on curved track at a certain speed undergoes a centrifugal lateral acceleration which can result in a lot of undesirable effects such as passenger discomfort, risk of vehicle overturning in combination with strong cross wind, derailment and so on [134] [80]. The most common track geometry parameters which are of great importance to railway vehicle dynamics are track gauge, track cant, transition curve radius, horizontal curve radius and vertical curve radius as well [85].

In summary, most of the foregoing studies on crosswind stability of railway vehicles are based on stationary wind scenarios, i.e. a deterministic wind speed variation with stationary wind turbulence. Vehicles with varying speed under nonstationary crosswind are not taken into account. Studies on the nonstationary wind process, especially on the vehicles running under nonstationary crosswind excitation are lacking. In addition, railway vehicles running on curved track under nonstationary wind excitation under different conditions (e.g. wind from outer or inner side) would also be of interest.

1.2.3 Road vehicles under crosswind

The response of road vehicles under nonstationary crosswinds is much more complicated. On the one hand, influences of crosswind on different types of vehicles are quite different. According to the foregoing research, lorries (especially high-sided lorries), vans and buses are most affected by strong winds, cf. [10] [45] [17]. Based on the collected data for wind-induced vehicle accidents that occurred in the UK during the storm in 1990 [17], 66% of the wind induced accidents involved vans and high-sided lorries while only 27% involved cars. Among these accidents, overturning is the most common failure mode, which amounts to about 47%. On the other hand, the state of road vehicles is more flexible due to the influence of the driver. An experienced driver may make the vehicle more stable when driving in a strong crosswind.

Baker [7] analyzed the crosswind stability of vehicles under the excitation of an artificial crosswind in a stationary process. A simplified vehicle model with single mass and four wheels running on a straight road was modeled. Three failure modes, namely, overturning, side slip and rotation, are considered in the study. Later, Baker [10] extended the theory by introducing a simple driver model and applied it to the investigation of a high-sided articulated road vehicle running under strong crosswind conditions. Results showed that the behavior of the driver had influence on crosswind stability. Besides, by parametric study, the most sensitive parameters were determined. It has been found out that vehicle parameters (e.g. mass and CG height), aerodynamic coefficients (rotational and lateral direction) and wind speed as well as road conditions have a significant effect on the crosswind stability of vehicles.

Recent publications for crosswind stability on road vehicles are [205] [186] and [45]. Chen and Cai [45] [32] presented a framework for crosswind analysis of the vehicle running on a long bridge in windy environment. Dynamic interaction between the vehicle and the bridge has been analyzed. The developed framework can be used not only for performance analysis of vehicle running on highways or bridges, but also in assessing the critical wind speed for the traffic so that potential accidents can be avoided. Snæbjörnsson [186] used a probabilistic model to assess the vehicle safety under strong crosswind based on reliability approach. The basic variables such as wind velocity, wind angle, vehicle speed, road camber and the frictional coefficient were modeled as independent stochastic variables. It has been suggested that the available probabilistic method as well as the reliability theory were valuable and can be well used for the risk analysis of crosswind stability of road vehicles.

As a whole, many investigations regarding crosswind stability of road vehicles have been carried out during the past decades. However, most of them were related to a deterministic wind model for a particular type of vehicle. Although the nonlinear aerodynamic coefficients are considered in some of the investigations, the stochastic wind model, including a gust model and turbulence, are usually simplified and modeled as stationary processes.

In summary, due to the existence of many uncertainties, a risk analysis has to be carried out and failure probabilities must be computed. The more realistic vehicle state, including a driver model and a nonstationary wind model as well as the different wind scenarios need to be investigated in a further step, so that a more exact assessment for the failure risk of road vehicles under nonstationary crosswind can be outlined. Moreover, road vehicles have to be classified into different categories since that the influences of crosswind acting on different vehicles are different.

1.2.4 Wind tunnel tests for vehicles under crosswind

To measure the aerodynamic forces acting on the vehicle, wind tunnel tests on scaled vehicle models are usually carried out. The most common research activities regarding wind tunnel tests for railway vehicles are related to drag reduction, crosswind stability, slip stream, aerodynamic noise and thermal aspects [15].

Until now, many investigations have been carried out on wind tunnel tests, cf. [8] [16] [190] [126] and [37]. One of the simplest ground scenarios that are widely used in the wind tunnel test is the 'flat ground' (FG) since it represents the basic configuration of ground, as has been specified in the national standard on crosswind effects on vehicles [15]. Conventional static low-turbulence wind tunnel experiments on train models cannot correctly reflect the real effect of the atmospheric turbulence or the relative motion between the moving vehicle and the ground. To overcome this drawback, Baker [8] carried out an improved experiment with a 1/50 advanced passenger train model which has been catapulted across the wind tunnel. An atmospheric boundary layer has been simulated by placing gravel chippings and a set of "spires" on the board. Moreover, the influence of wind fences on the aerodynamic coefficients has also been studied in this experiment. Recently, based on pressure field measurements, Sanquer [174] developed a new methodology in order to get the local and instantaneous aerodynamic loads. The proposed method allows for a global understanding of the distribution of aerodynamic loads along the vehicle and gives a new way to assess the aerodynamic forces caused by crosswind.

Besides, Bocciolone [29] performed an experimental test on three types of railway vehicles in different configurations. This test allowed to perform a sensitivity analysis on the main parameters that influence the aerodynamic behavior of the vehicle. The effects of the train motion and turbulence intensity on the aerodynamic coefficients were studied. For a static train, the aerodynamic coefficients (especially for the higher angles of attack) are significantly influenced by the wind turbulence. For a moving vehicle, results showed that there was no significant influence of the vehicle motion on the aerodynamic coefficients. However, this conclusion needs to be further verified.

Moreover, wind tunnel tests for road vehicles have also been carried out in the past decades, e.g. [56] and [65]. A specific task of the EC project WEATHER [65] has been devoted to the research of developing a wind alarm system for road vehicles to reduce the rollover risk associated to strong crosswinds. To achieve this aim, various investigations have been carried out, e.g. CFD simulations, wind tunnel experiments and full scale tests [38]. A detailed comparison of these methods for the evaluation of the wind force acting on a high sided lorry can be seen in [189].

In order to study the flow around a road vehicle running in strong crosswind, Coleman and Baker [56] carried out a wind tunnel test on a 1/50th scale articulated lorry model on the bridge deck with an oil-flow-visualization test. The mean and fluctuating values of aerodynamic force and moment coefficients were measured. Studies showed that fluctuations of the side force were essentially caused by turbulence buffeting while lift force fluctuations were induced by wake vortex shedding. Similar results have also been reported in [11] and [55].

In recent wind tunnel tests for heavy road vehicles [38] [40], the aerodynamic forces and moments are obtained and compared under different testing parameters, including vehicle types, infrastructure scenarios (flat ground, embankment, single and double viaduct) and turbulence conditions. It has been found out that the main influence of the infrastructure scenario is on the lateral force and roll moment coefficients especially when the yaw angle is high. In addition, investigations indicated that the vehicle cross section geometry has a significant influence on the steady part of the lateral aerodynamic force but not on the unsteady part. Larger vehicle side areas exposed to the wind will make the lateral force admittance function decrease more.

Although a number of wind tunnel experiments have been carried out in the past, the determination of aerodynamic coefficients for different vehicles in various realistic wind scenarios is still a challenge. To verify the experimental data, more tests have to be carried out. In addition, it is expensive and very hard to determine the aerodynamic coefficients for each type of vehicle in every wind scenario. Therefore, numerical simulations as well as full scale tests are necessary.

1.2.5 Risk and sensitivity analysis

A risk analysis can identify the hazard probability of vehicles running in various wind scenarios. So far, risk assessment procedures for crosswind related accidents proposed in the literature can be basically divided into two groups [161] [15].

The first group of methods computes critical wind speeds for the vehicle with deterministic values that characterizes a failure condition such as the vehicle speed, the wind angle, and the height of the wind pressure center and so on. In this regard, the risk assessment procedure can be divided into two steps:

- Simulating the vehicle dynamics under strong crosswind with a given wind profile so that the critical wind speed can be determined.

- Assessing the risk by evaluating the occurrence of wind speed exceeding the critical wind speed.

The underlying assumption for this method is that the critical wind speed can be calculated independently under the condition that a deterministic gust shape is predetermined. Cooper [58] proposed a risk assessment based on the assumption that the ratio of the wind speed to a reference speed exceeding the critical value (i.e. the exceeding probability) obeys a Gumbel distribution. The attack angle of wind is taken into consideration, and sensitivities in terms of the overturning probability under various parameters are stated. A similar method was also proposed in [72], in which the statistical distribution of characteristics wind curve (CWC) has been also considered.

The second group takes uncertainties into account. As a result, all the deterministic parameters (e.g. aerodynamic coefficients, gust amplitude and duration) are substituted by stochastic variables. As far as the author knows, the earliest work concerning the crosswind problem that has applied the stochastic method for the design of wind loaded structures is [63]. Because of lack of knowledge, the related random variables are usually assumed to obey a normal distribution [202], which may need to be checked and analyzed for practical use.

Carrarini [34] investigated the influence of uncertainties by considering both the aerodynamic part and vehicle dynamic part. He assumes an extreme value distribution for the average wind velocity at the site. The corresponding failure probability for the vehicle as a function of mean wind speed has been calculated. Furthermore, Wetzel [204] carried out a reliability analysis and a sensitivity analysis for both railway and road vehicles excited by strong crosswinds. Wind uncertainties as well as vehicle uncertainties (e.g. vehicle mass, height of gravity center) were taken into account.

Generally speaking, the risk of vehicles running under strong crosswind is usually determined by analytical methods [167], semi-analytical methods [136] and numerical methods [181]. The analytical methods determine the failure probability of a vehicle by an extreme value analysis. In this aspect, the maximum method (MAX) and peaks over threshold method (POT) are usually carried out. On the other hand, semi-analytical methods (e.g. first or second order reliability method) are widely used nowadays because they can efficiently handle problems with a relatively low number of random variables. A general first step for the semi-analytical method is to perform a transformation of all random variables to standard normal distributions. This can be done by means of Rosenblatt [169] or Nataf [148] transformation. Moreover, the numerical methods for failure probability evaluations are mainly based on Monte-Carlo simulation (MCS) and its variants.

A disadvantage for the traditional MCS method is that it needs a lot of independent samples in order to achieve a sufficient accuracy. To circumvent this problem, many other efficient methods are used such as importance sampling (IS) and line sampling (LS). The idea of importance sampling is to alter the sampling procedure in such a way that more samplings lie in the failure domain. The line sampling method is very efficient if a good important direction can be found. Otherwise the computational costs will increase drastically and the results can also be biased, cf. [205], [123] and [176].

In order to investigate the influence of input parameters on the output of a system, a sensitivity analysis is usually carried out. Sensitivity analysis procedures can be classified into two kinds, the local and global methods, respectively. In general, the local method which represents less information about the system is based on a gradient at certain nominal values while the global method yields the average values over the whole parameter space. Compared to the local method, the global method is numerically more expensive.

For a stochastic system, the local influence of random variables on the failure probability can be computed by a local semi-analytical method. It gives the gradient value of the variable at the most probable point (MPP), see [15] [205]. When global sensitivity analysis is considered, the commonly used methods like Morris approach and correlation analysis [205] [173] are usually applied. In comparison to correlation analysis, the Morris method is in general numerically less demanding but unfortunately it yields only qualitative results [173].

1.3 Objective of the work

The purpose of this dissertation is to study the crosswind stability of vehicles under nonstationary wind excitation. Ground vehicles, including railway vehicle and road vehicle, running in a nonstationary scenario under various different conditions (e.g. different vehicle speeds, dry/wet road, different track parameters) will be considered.

To begin with, the nonstationary character of natural wind speeds will be studied. Different methods for producing nonstationary wind turbulence will be discussed. On this basis, a stochastic gust model with nonstationary wind turbulence will be proposed for the risk analysis.

Then, risk assessment based on reliability analysis will be carried out for both railway and road vehicles. For railway vehicles, crosswind stability of a vehicle running on straight and curved tracks will be studied. For road vehicles, due to the fact that the influence of crosswind on different types of vehicles is very different, models for different vehicles (e.g. trucks, cars, car-trailers and etc.) have to be developed. Besides, for each type of vehicle, the corresponding aerodynamic coefficients, which are highly dependent on vehicle shapes and usually fitted by the data from a limited number of wind tunnel experiments, have to be identified. Moreover, failure criteria for different vehicles need to be defined, that is, overturning for railway vehicles and overturning as well as slide slip and yaw slip for road vehicles.

Furthermore, this work will also study some special aspects of the crosswind stability of vehicles. These aspects are of great importance but were usually simplified or not given so much attention in the foregoing research. For example, trains running on curved track under nonstationary wind excitation and trains running with varying velocity (acceleration/deceleration) are considered. For road vehicles, in order to evaluate the failure probability in a more realistic wind scenario, on site measured wind data will be analyzed and utilized in the simulation.

In summary, based on the nonstationary crosswind as well as the nonstationary vehicle state, a combined vehicle-wind system will be modeled and simulated. As the excitation of the vehicle is a stochastic process, a risk analysis will be carried out and failure probabilities will be computed. Railway vehicles with nonstationary wind excitation on straight/curved tracks and road vehicles with wind excitation under various conditions will be taken into account. The nonstationary aerodynamic force produced by the nonstationary crosswind excitation will be studied and analyzed in detail. Sensitivity analysis, which is an efficient way to investigate the influence of input-parameters on the output of the system, will also be carried out. The effect of countermeasures such as vehicle speed reduction and wind fences can be judged on an objective basis.

1.4 Structure of the work

The dissertation is organized into eight chapters and structured in the order of nonstationary wind models, nonstationary wind excitations, vehicle models and risk analysis.

This chapter introduces the background and motivation of the research. Some important aspects regarding nonstationary crosswind stability of ground vehicles are reviewed. The research methods as well as the objectives of the thesis are presented.

Next, in Chapter 2, the probabilistic characteristics of the wind speed are studied. Natural wind is a stochastic process. To produce the turbulent wind, different methods (e.g. PSD, wavelet method, NONAR method) are described. After that, a stochastic gust model with the nonstationary turbulence based on NONAR method is proposed.

Chapter 3 analyzes the wind force excited by the nonstationary wind that acts on the moving vehicles. For turbulent wind forces, the aerodynamic admittance function is studied. In addition, special phenomena like vortex shedding and buffeting for railway vehicles have also been discussed. The Hilbert-Huang transform (HHT) is carried out for the nonstationary analysis.

Chapter 4 details the modeling and simulation for both road and railway vehicles. On the one hand, the road vehicles are divided into several types. For each vehicle, the corresponding vehicle model will be identified. Driver model and road model are also taken into account. On the other hand, railway vehicles running on straight and curved tracks are investigated. For vehicles running on straight track, the influence of track irregularity on wheel/rail contact has been discussed. When the vehicle runs on curved track, the unbalanced lateral acceleration will be studied.

Chapter 5 represents the probabilistic method for risk analysis of crosswind stability of vehicles running under strong crosswind. Failure criteria for both railway and road vehicles due to strong crosswind are clarified and failure probabilities under different conditions are defined. FORM and Line sampling (LS) are utilized as the main methods for reliability analysis.

In Chapter 6, risk analysis for railway vehicles excited by nonstationary strong crosswind have been carried out. A stochastic gust model with nonstationary wind turbulence as well as a nonstationary vehicle state (acceleration/deceleration) has been considered. Failure probabilities for a vehicle running in different scenarios (e.g. straight track, curved track etc.) are computed. In addition, a vehicle running through an artificial storm in a nonstationary scenario has also been studied.

Chapter 7 investigates the crosswind stability of road vehicles running under strong crosswinds in various scenarios (e.g. dry road, wet road, different wind angles). Different types of vehicles are taken into account and the measured wind data are fitted for the stochastic gust model. For each type of vehicle, risk analysis is carried out and failure probabilities are computed. Based on the results of reliability analysis as well as the actual data of the wind-induced accidents in the past, the vehicle speed limit or the corresponding parameters for a wind warning system can be proposed.

Finally, Chapter 8 summarizes the main results and contributions of this work. Recommendations for further studies are also presented.

The main contents and a general structure of this dissertation can be seen in Figure 1.2.

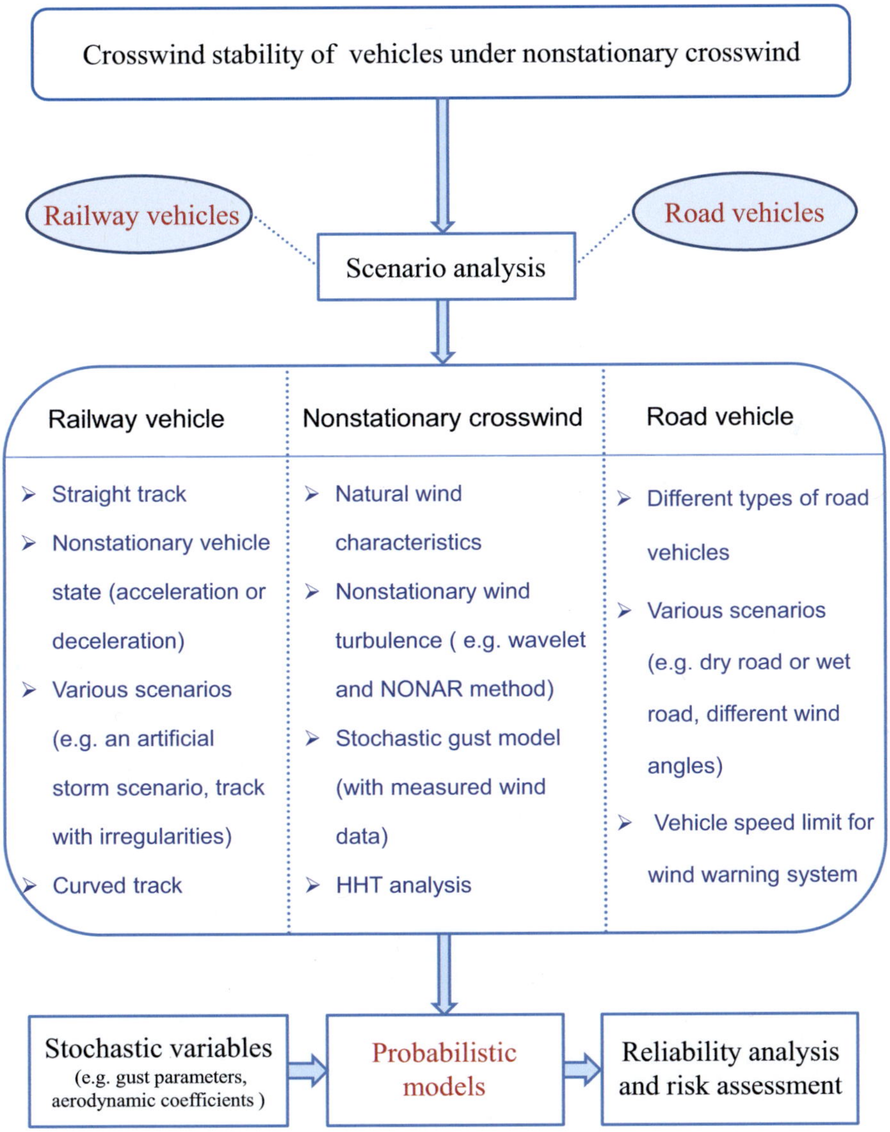

Figure 1.2: A general structure of the main contents in this dissertation

2 Nonstationary wind

This chapter starts with the fundamental theory of natural wind, followed by a section on the probabilistic characteristics of the wind speed. The distribution of natural wind speeds is studied. In the next section, different methods for producing the nonstationary wind turbulence are discussed. Finally, a stochastic gust model with nonstationary wind turbulence based on NONAR method is presented.

Global winds are caused by pressure differences on the earth surface due to the uneven heating by solar radiation [111] [139]. They are also affected by the topography like large oceans, high mountains and so on. Analysis of wind characteristics in the foregoing research is usually based on a stationary model [41], in which the wind turbulence is assumed as an ergodic stochastic process. However, recent studies have shown that wind velocity as well as its effects on buildings and structures are not always stationary, this can be seen in [132] [92] [118] [117] and [212]. Kareem [117] pointed out that most extreme wind events are nonstationary and often highly transient, such as downbursts and hurricanes. Stochastic winds with instantaneous changes in their speeds would potentially affect a structure. More efforts in studying the nonstationary features of winds in transient events can be found in [199] [42] and [41].

In order to study the crosswind stability of vehicles in a more realistic way, physical assumptions on the nature of crosswind have to take its nonstationary character into consideration.

On the whole, the statistical wind with turbulence is usually characterized by the following aspects [139]:

a) Probability distributions of wind speed

b) Turbulence intensity and integral length

c) Power spectral density function

d) Autocorrelation function.

For a certain point at height z in space, the nonstationary wind speed $U(z,t)$ is usually considered to be composed of three components and can be described as follows [211]:

$$U(z,t) = \begin{pmatrix} u_u(z,t) \\ u_v(z,t) \\ u_w(z,t) \end{pmatrix} = \begin{pmatrix} u_0(z) + u'_u(z,t) \\ u'_v(z,t) \\ u'_w(z,t) \end{pmatrix} \tag{2.1}$$

where $u_u(z,t)$, $u_v(z,t)$, $u_w(z,t)$ refer to the three components of wind speed in longitudinal, lateral and vertical direction, respectively. $u_0(z)$ is the mean wind speed and $u'_u(z,t)$, $u'_v(z,t)$, $u'_w(z,t)$ are the corresponding wind turbulence in each direction.

Compared to the other components, the wind speed in mean wind direction (here, i.e. the longitudinal direction) is usually considered as the main wind speed. It is often observed in time series and measured from an anemometer [139] [202]. In this dissertation, unless specified, only wind speeds in the mean wind direction are considered. Following this, the total wind speed $U(z,t)$ can be approximated as

$$U(z,t) \approx u(z,t) = u_0(z) + u'(z,t) \tag{2.2}$$

where $u(z,t)$, $u'(z,t)$ are the abbreviations for $u_u(z,t)$, $u'_u(z,t)$, respectively.

For a fixed height z, the wind speed only changes with time. Thus, Equation (2.2) can be simplified to

$$u(t) = u_0 + u'(t) \tag{2.3}$$

To estimate the approximate mean wind speed u_0 at a referent height in space, the well-known logarithmic wind profile [139] is usually applied:

$$u_0(z) = \frac{u^*}{k} \ln(\frac{z+z_0}{z_0}) \tag{2.4}$$

where u^* is the friction velocity, k is a constant, z_0 refers to the surface roughness length. A summary of the values for z_0 with different surface types can be seen in Table 2.1.

Figure 2.1 shows an example of the the real wind data measured at two different places on two days, respectively. The wind speeds are measured every one minute at place 1 and every two minutes at place 2. Corresponding distribution of the wind directions can be seen in Figure 2.2. As can be seen from the figure, the main wind direction at place 1 on the first day are between $157° \sim 180°$ and $270° \sim 292.5°$ while on the second day are between $0° \sim 135°$ and $292.5° \sim 360°$. The wind speed as well as its direction are changing with time at both places. Therefore, risk analysis for vehicles under strong crosswind has to take both the wind speed and direction into consideration. Figure 2.3 represents the mean and maximum values of wind speeds in different directions, from which the largest wind speed with its corresponding wind direction can be determined.

Table 2.1: Roughness length z_0 of homogeneous surface types [207]

Surface type	Roughness length (m)
Sea, loose sand and snow	0.0002 (U-dependent)
Concrete, flat desert, tidal flat	0.0002 – 0.0005
Flat snow field	0.0001 – 0.0007
Rough ice field	0.001 – 0.012
Fallow ground	0.001 – 0.004
Short grass and moss	0.008 – 0.03
Long grass and heather	0.02 – 0.06
Low mature agricultural crops	0.04 – 0.09
High mature crops ("grain")	0.12 – 0.18
Continuous bushland	0.35 – 0.45
Mature pine forest	0.8 – 1.6
Tropical forest	1.7 – 2.3
Dense low buildings ("suburb")	0.4 – 0.7
Regularly-built large town	0.7 – 1.5

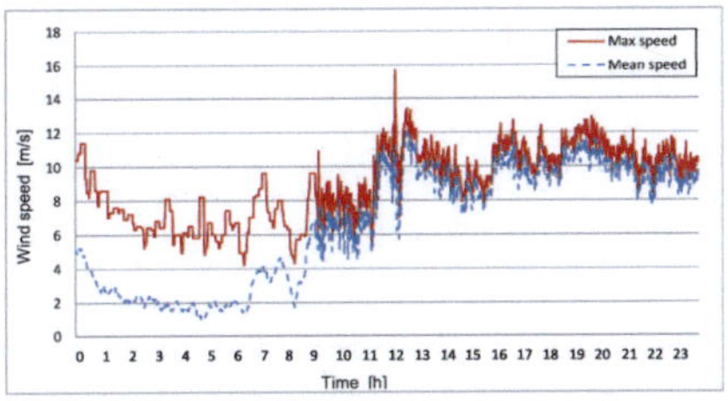

a) place 1, day 1

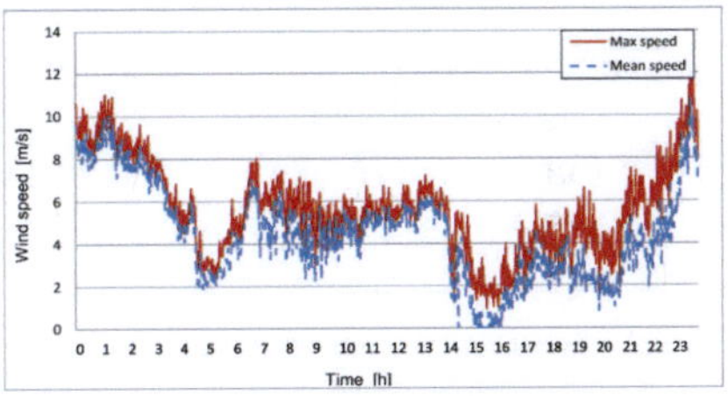

b) place 1, day 2

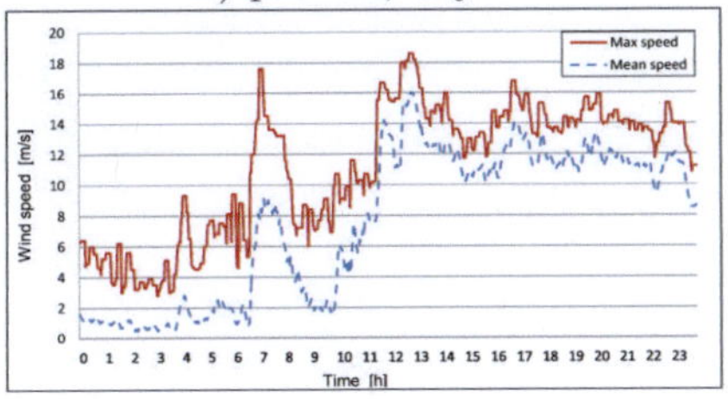

c) place 2, day 1

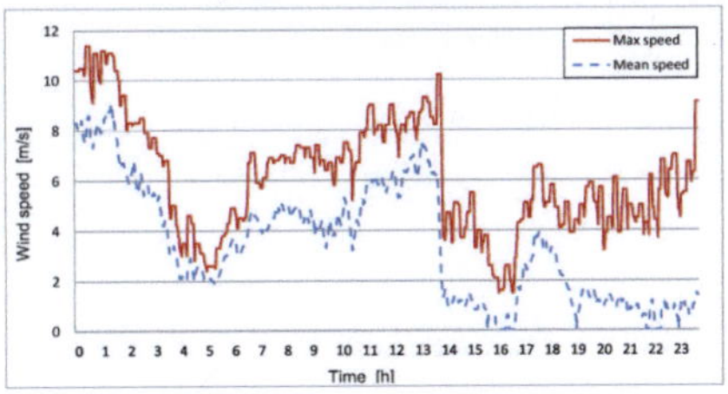

d) place 2, day 2

Figure 2.1: An example of the real wind data measured from two different places on two days

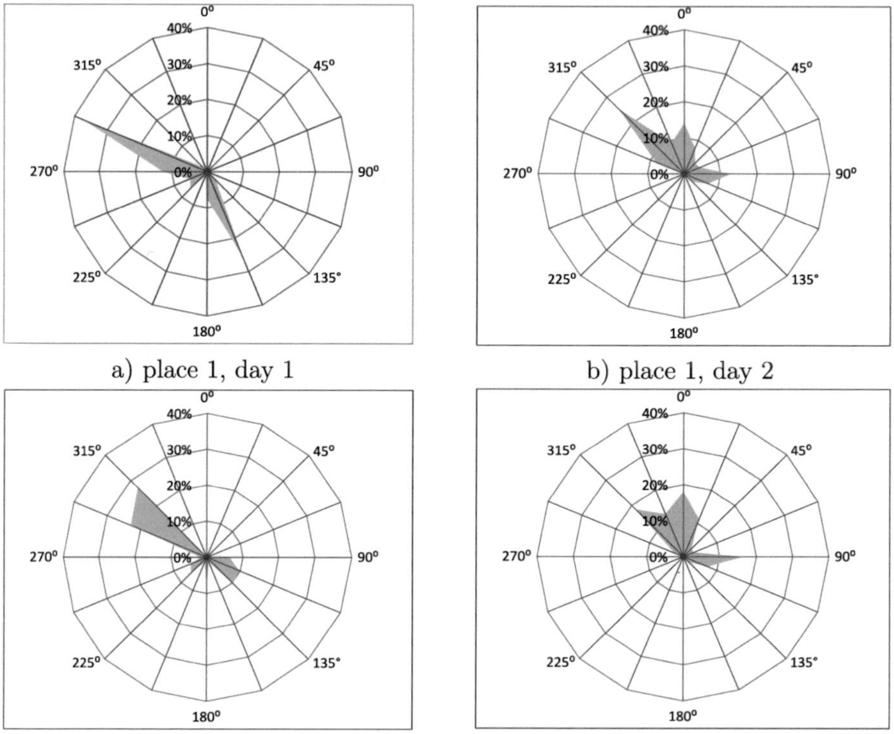

a) place 1, day 1

b) place 1, day 2

c) place 2, day 1

d) place 2, day 2

Figure 2.2: The distribution of wind directions for actually measured wind data

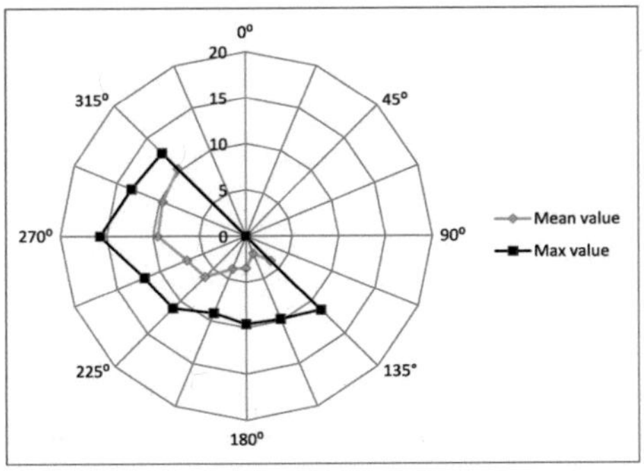

a) place 1, day 1

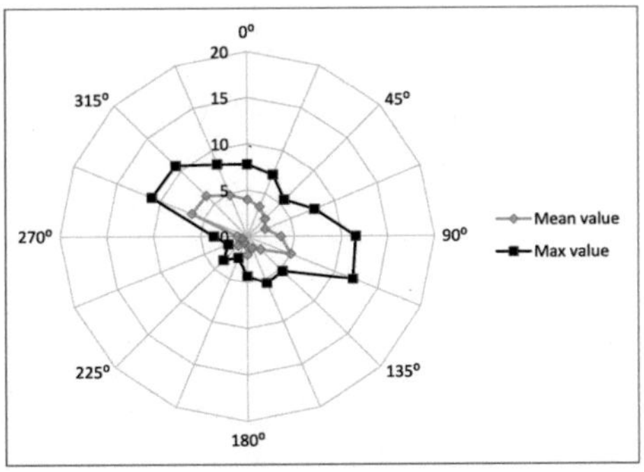

b) place 1, day 2

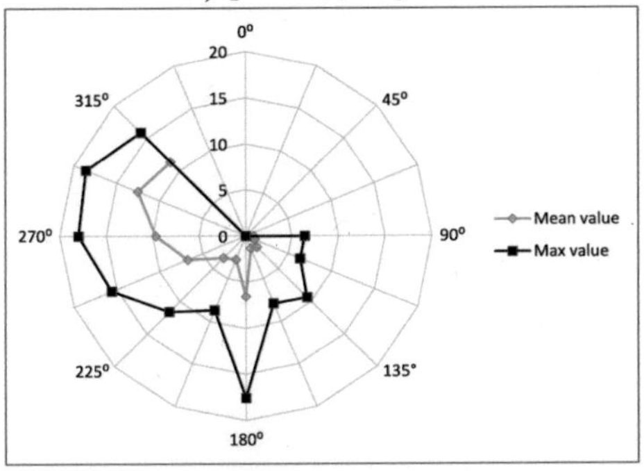

c) place 2, day 1

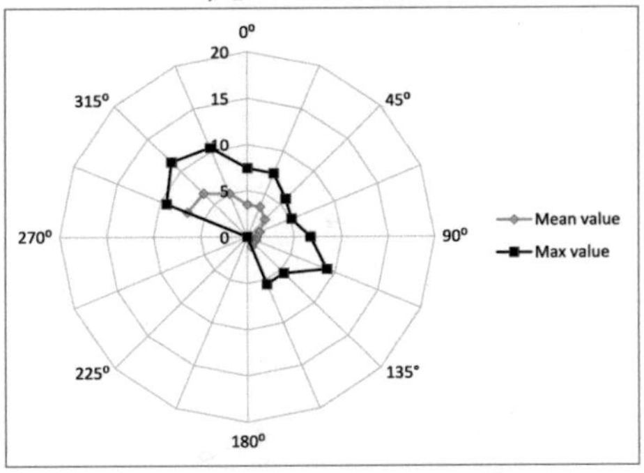

d) place 2, day 2

Figure 2.3: Mean and max values of wind speeds in different directions

2.1 Basic characteristics of natural wind

Due to the existence of turbulence, actual measurements of the wind speed usually last for a long time so that the mean wind speed as well as other statistical characteristics of the wind speed can be obtained.

Consider a set of wind data $\{u_1, u_2, \cdots u_N\}$ measured from a real wind field at the same position with a constant time step. The mean wind speed can be estimated as follows

$$u_0 = \frac{1}{N}\sum_{i=1}^{N} u_i \tag{2.5}$$

Based on u_0, the standard deviation of the wind speed can be estimated from

$$\sigma_u = \sqrt{\frac{1}{N-1}\sum_{i=1}^{N} (u_i - u_0)^2} \tag{2.6}$$

2.2 Probability density function of the wind speed

In order to analyze the frequency of actual measured wind data, the measured wind data are usually divided into several small groups with $u_j \leq u < u_{j+1}$, $j = 1, 2, \cdots n$, where n represents the total number of wind speed intervals. Then the frequency for the wind speed in each interval can be calculated by

$$f_j = \frac{N_j}{N} \tag{2.7}$$

in which f_j refers to the frequency and N_j is the total number of wind speed data falling in the interval $[u_j, u_{j+1})$.

In practice, the distribution of wind speeds is often described by a probability density function (PDF). For example, the Weibull PDF and its special case, Rayleigh PDF are usually used for fitting the PDF of wind data. They describe a relatively likelihood for the wind speed at a certain value.

For a given interval $[u_a, u_b]$, the probability of the wind speed u lying in this range can be obtained by

$$P(u_a \leq u \leq u_b) = \int_{u_a}^{u_b} f(u) d_u = F_u(u_b) - F(u_a) \tag{2.8}$$

where $f(u)$ refers to the PDF of u and F_u is the corresponding cumulative distribution function (CDF). For a certain wind speed U_0, $F_u(U_0)$ can be defined as

$$F_u(U_0) = P(u \leq U_0) = \int_0^{U_0} f(u) d_u \tag{2.9}$$

As a typical distribution of the wind speed, Weibull distribution [76] has been widely used in wind engineering. It has been proved that the Weibull PDF provides a convenient representation for the wind speed and gives a good fit to the actual measured wind data [177] [114]. According to [175], the Weibull PDF with three parameters can be expressed as follows:

$$f(u) = \frac{k}{b-a}\left(\frac{u-a}{b-a}\right)^{k-1} \exp\left[-\left(\frac{u-a}{b-a}\right)^{k}\right] \tag{2.10}$$

where k, a and b refer to the non-dimensional shape parameter, location parameter and scale parameter, respectively. The location parameter a is equal to the minimum value of u, i.e. $u \geq a$.

Since the minimum wind speed is 0, the Weibull PDF is usually simplified as the two-parameter function:

$$f(u) = \frac{k}{b}\left(\frac{u}{b}\right)^{k-1} \exp\left[-\left(\frac{u}{b}\right)^{k}\right] \tag{2.11}$$

The corresponding cumulative distribution function can be derived as follows:

$$F_u(U_0) = P(u \leq U_0) = \int_0^{U_0} f(u)\mathrm{d}u = \int_0^{U_0} \frac{k}{b}\left(\frac{u}{b}\right)^{k-1} \exp\left[-\left(\frac{u}{b}\right)^{k}\right] \mathrm{d}u = 1 - e^{-(U_0/b)^k} \tag{2.12}$$

Rayleigh distribution is actually a special case of the Weibull distribution with $k = 2$. In this case, the probability density function can be described as follows:

$$f(u) = \frac{2u}{b^2} \cdot e^{-\left(\frac{u}{b}\right)^2} \tag{2.13}$$

Accordingly, the cumulative distribution function for the Rayleigh distribution can be defined as

$$F_u(U_0) = P(u \leq U_0) = 1 - e^{-(U_0/b)^2} \tag{2.14}$$

The mean value and variance of the Weibull distribution and Rayleigh distribution are given by [133]

$$\mu_W = b\Gamma(\frac{1}{k}+1) \ , \ \ \mu_R = \frac{\sqrt{\pi}}{2}b \tag{2.15}$$

$$\sigma_W^2 = b^2\left[\Gamma\left(\frac{2}{k}+1\right) - \Gamma^2\left(\frac{1}{k}+1\right)\right] \ , \ \ \sigma_R^2 = b^2\left(2-\frac{\pi}{4}\right) \tag{2.16}$$

where the subscript 'W' and 'R' refer to the Weibull distribution and Rayleigh distribution, respectively. $\Gamma(t)$ is the gamma function which can be expressed with

$$\Gamma(t) = \int_0^\infty x^{t-1}e^{-x}\mathrm{d}x \tag{2.17}$$

When real wind data are fitted to the commonly used two-parameter Weibull distribution, the two unknown parameters, i.e. the shape parameter k and scale parameter b have to be estimated. So far, there are many mathematical methods for estimating the Weibull parameters, such as the least squares method, the moment method, the graphical method and the maximum likelihood method and so on, cf.[114] [76] and [177]. In this work, the maximum likelihood method which has been recommended in [177] is applied. Based on the maximum likelihood method, the parameters k and b can be well determined by the following equations.

$$k = \left(\frac{\sum\limits_{j=1}^{N} u_j^k \ln(u_j)}{\sum\limits_{j=1}^{N} u_j^k} - \frac{1}{N}\sum_{j=1}^{N}\ln(u_j) \right)^{-1} \tag{2.18}$$

$$b = \left(\frac{1}{N}\sum_{j=1}^{N} u_i^k\right)^{\frac{1}{k}} \tag{2.19}$$

It should be paid attention that in the calculation of parameter k, only the nonzero wind speed can be applied. In practice, wind speeds at 0 m/s are usually replaced by a very small number, e.g. 0.001 m/s. Figure 2.4 and 2.5 represent the properties of Weibull PDF with different parameter values. As can be seen in Figure 2.4, the maximum value of PDF becomes larger when the parameter k increases. For a special case, i.e. $k = 1$, the Weibull distribution becomes an exponential distribution. In Figure 2.5, the shape of PDF for

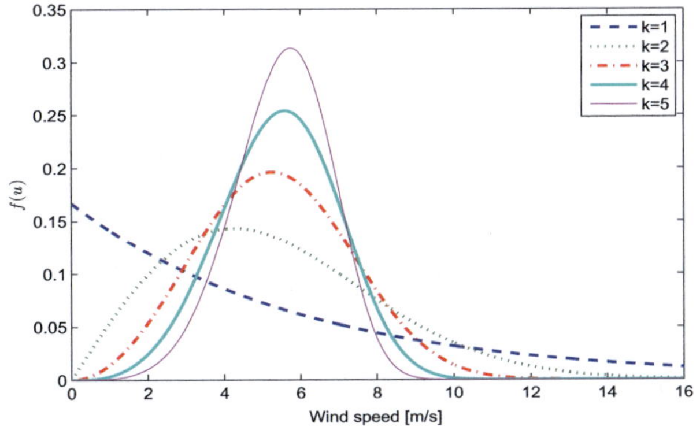

Figure 2.4: Probability density function for Weibull distribution at $b = 6$ with different k

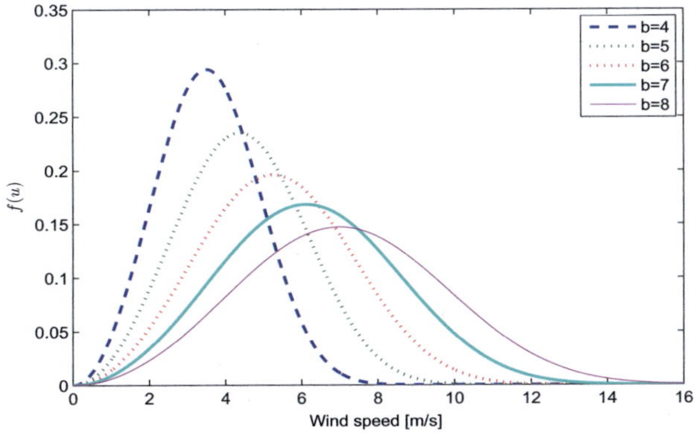

Figure 2.5: Probability density function for Weibull distribution at $k = 3$ with different b

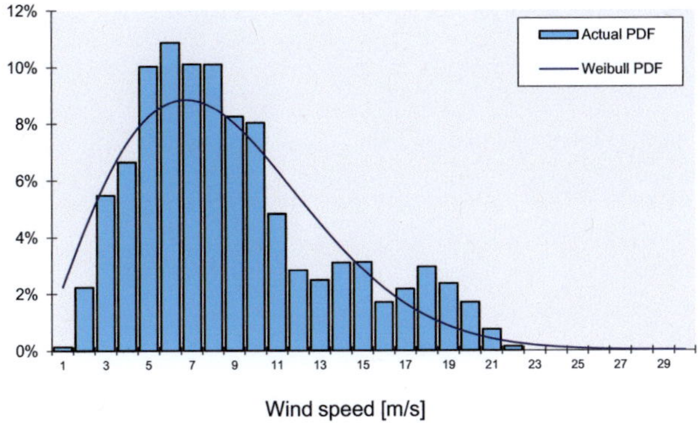

Figure 2.6: An example of the PDF for the measured wind speed

Weibull distribution gets wider with increasing values for parameter b. Both figures show the important characteristics of Weibull distribution. Based on the conclusion described above, the trend for the fitting of actual wind speed data can be well evaluated.

Figure 2.6 shows an example of the PDF for wind speeds measured every one minute during a month at a certain place. As is shown, the actual PDF of wind speeds coincides with the Weibull PDF well.

2.3 Wind turbulence

As mentioned before, a lot of studies have been carried out to investigate the wind modeling for wind-structure interactions over the last few decades. The main theory, however, is based on a stationary Gaussian process [92]. To be more realistic, the simulation of natural crosswind has to take its nonstationary characteristics into consideration.

To simulate the nonstationary turbulence, several methods have been proposed so far, such as the wavelet method, evolutionary power spectral density (EPSD) method and ARMA, cf. [88] [147] [132] [197] [157]. Among these methods, the wavelet and ARMA methods have attracted a lot of attention. Examples for the wavelet method can be seen in [122] [92]. Compared to the wavelet method, the ARMA method has been proved to be the most efficient in terms of saving computational effort [149]. Rossi [170] gives a detailed description for ARMA modeling and evaluates the performance of various numerical procedures for AR and ARMA, while Maeda and Makino [137] use ARMA to simulate the turbulence of gust wind.

2.3.1 Power spectral decomposition

Traditional methods for the simulation of wind turbulence are based on the decomposition of power spectral density (PSD). Following this method, a stochastic process can be simulated by the following series with $N \to \infty$ according to [182] [71]

$$u'(t) = \sum_{i=0}^{N-1} \sqrt{S_{u'u'}(f_i) \cdot \Delta f} \cdot \cos(2\pi f_i t + \varphi_i) \tag{2.20}$$

with $f_i = i \cdot \Delta f$, $\Delta f = f_c/N$,

in which, f_c refers to the upper cut-off frequency, φ_i are independent and identically distributed (i.i.d.) random phase angles. They obey the uniform distribution over the interval $[0, 2\pi]$. $S_{u'u'}(f)$ is the wind power spectrum.

Table 2.2 presents the main wind spectra that have been used in the past. They can be generally classified into two groups, namely, the height-independent and height-dependent group [21]. Among these spectra, the von Kármán spectrum has been found to be a satisfactory model in a variety of contexts [86]. It provides very good approximations to the wind spectra and has been widely used, e.g. [112] [86].

Based on Equation (2.20), stationary, Gaussian stochastic wind processes can be simulated. However, it should be noted that the traditional method based on PSD cannot be used directly for the simulation of nonstationary turbulence. To settle this problem, other corresponding methods have to be carried out.

Table 2.2: Different types of wind spectra [21] [200] [184] [145]

Height independent	Harris spectrum	$\frac{fS_{u'u'}(f)}{u^*} = \frac{4x^2}{(1+x^2)^{5/6}}$	$x = \frac{1800f}{u_0}$
	Davenport spectrum	$\frac{fS_{u'u'}(f)}{u^*} = \frac{4x^2}{(1+x^2)^{4/3}}$	$x = \frac{1200f}{u_0}$
Height dependent	Simiu spectrum	$\frac{fS_{u'u'}(f)}{u^*} = \frac{200\tilde{f}}{\left(1+50\tilde{f}\right)^{5/3}}$	$\tilde{f} = \frac{fz}{u_0}$
	Kaimal spectrum	$\frac{fS_{u'u'}(f)}{u^*} = \frac{105\tilde{f}}{\left(1+33\tilde{f}\right)^{5/3}}$	$\tilde{f} = \frac{fz}{u_0}$
	von Kármán spectrum	$\frac{fS_{u'u'}(f)}{\sigma_{u'}^2} = \frac{4\tilde{f}}{\left(1+70.8\tilde{f}^2\right)^{5/6}}$	$\tilde{f} = \frac{fL_{u'}}{u_0}$

2.3.2 Evolutionary power spectrum density

Assume that the fluctuating wind speed $u'(z,t)$ is a uniformly modulated evolutionary process, which can be modeled as a zero mean stationary process modulated with a deterministic modulation function [47] [43], i.e.:

$$u'(z,t) = a(z,t) \cdot \Psi(z,t) \tag{2.21}$$

where $a(z,t)$ refers to the modulating function. $\Psi(z,t)$ is a stationary Gaussian stochastic process, which can be described as

$$\Psi(z,t) = \int_{-\infty}^{+\infty} e^{i2\pi ft} \mathrm{d}_{\mathbb{Z}(z,f)} \tag{2.22}$$

$$E\{\mid \mathrm{d}_{\mathbb{Z}(z,f)} \mid^2\} = \Theta(z,f)\mathrm{d}_f \tag{2.23}$$

in which, f refers to the frequency, $\mathbb{Z}(z,f)$ is an orthogonal increment process and $\Theta(z,f)$ is the power spectral density of $\Psi(z,t)$. For any given z, $\int_{-\infty}^{+\infty} \Theta(z,f)\mathrm{d}_f = 1$.

The evolutionary power spectrum density (EPSD) of $u'(z,t)$ and the evolutionary cross-spectrum density for $u'(z_1,t)$ and $u'(z_2,t)$ are given as

$$S_{u'u'}(z,f,t) = \mid a(z,t) \mid^2 \Theta(z,f) \tag{2.24}$$

$$S_{u'u'}(z_1,z_2,f,t) = a(z_1,t)a^*(z_2,t)\Theta(z_1,z_2,f) \tag{2.25}$$

where '*' refers to the complex conjugate, $\Theta(z_1, z_2, f)$ is the cross-spectral density function between $u'(z_1, t)$ and $u'(z_2, t)$, which can be expressed by the product of power spectral densities and the coherence function $\mathrm{Coh}(f)$.

$$| \Theta(z_1, z_2, f) |^2 = \Theta(z_1, f)\Theta(z_{2,}f)\mathrm{Coh}(f) \tag{2.26}$$

Combining Equation (2.25) and (2.26), the evolutionary cross-spectrum density between $u'(z_1, t)$ and $u'(z_2, t)$ can be rewritten as

$$S_{u'u'}(z_1, z_2, f, t) = a(z_1, t)a^*(z_2, t)\sqrt{\Theta(z_1, f)\Theta(z_2, f)\mathrm{Coh}(f)} \tag{2.27}$$

For any two points $A(x_i, y_i, z_i)$, $B(x_j, y_j, z_j)$ in space, the wind coherence function can be approximated by an exponential function [214] [184]:

$$\mathrm{Coh}(f) = \exp(-\frac{2f\sqrt{C_x^2(x_i - x_j)^2 + C_y^2(y_i - y_j)^2 + C_z^2(z_i - z_j)^2}}{u_0(z_i) + u_0(z_j)}) \tag{2.28}$$

where $u_0(z_i)$, $u_0(z_j)$ are the mean wind speed at the height of z_i and z_j, respectively. C_x, C_y and C_z are the corresponding empirical decay constants.

Chen [43] utilized EPSD to model the stochastic fluctuation in down bursts. Based on the study of EPSD with full-scale and wind tunnel data, it has been found out that the deviation of the wind fluctuation appears to be about 1/4 of the mean wind speed and later Chen [36] [44] suggested that $a(z, t)$ is equal to 0.08 to 0.11 times the non-turbulent wind speed.

2.3.3 Wavelet method for the analysis of nonstationary turbulence

For several years, wavelet method has been employed to analyze the nonstationary wind speed, cf. [95], [1]. Although the traditional inverse Fourier transform (IFT) can be used to generate time series of a stochastic process, it is difficult to represent the time-dependent characteristics of a synthesized signal with IFT [122].

The wavelet method overcomes the limitation of Fourier analysis (FT). Unlike FT, it deals with expansion of functions in terms of basis wavelet functions but not trigonometric polynomials. In wavelet analysis, the wavelets are generated in the form of translations and dilations of a fixed function called the mother wavelet $\psi(t)$, which can be expressed as follows [62]:

$$\psi(t) = \frac{1}{\sqrt{a}}\psi(\frac{t-b}{a}), \quad a, b \in \mathrm{R}\ \mathrm{and}\ a \neq 0 \tag{2.29}$$

where a, b are the scale and translation parameters for the wavelet function, t refers to the time.

Based on the mother wavelet, the continuous wavelet transform (CWT) of a real signal $y(t)$ ($y(t) \in L^2(R)$, $L^2(R)$ is the space of real square summable functions [95]) with respect to $\psi(t)$ can be defined by:

$$w(a,b) = \langle y(t), \psi(t) \rangle = \frac{1}{\sqrt{a}} \int_{-\infty}^{\infty} \psi(\frac{t-b}{a}) y(t) \mathrm{d}t \tag{2.30}$$

Correspondingly, the inverse continuous wavelet transform (ICWT) of the signal is

$$y(t) = \frac{1}{C_\psi} \int_{-\infty}^{+\infty} \int_{-\infty}^{+\infty} w(a,b)\psi(t) \frac{1}{a^2} \mathrm{d}a \mathrm{d}b \tag{2.31}$$

In practice, it is usual to adopt the discrete-binary wavelet function [213], which can be expressed as follows [130]

$$\psi_{j,k}(t) = 2^{j/2} \psi(2^j t - k) \tag{2.32}$$

where $\psi_{j,k}(t)$ are the wavelet functions, j is the scale parameter, k is the translation parameter, $j, k \in z$.

Corresponding discrete wavelet transform (DWT) of $y(t)$ can be obtained by the following equation

$$w_{j,k} = \int_{-\infty}^{+\infty} y(t) \psi_{j,k}^*(t) \mathrm{d}t \tag{2.33}$$

where '*' means the complex conjugate. The inverse discrete wavelet transform (IDWT) of the signal is

$$y(t) = \sum_j \sum_k w_{j,k} \psi_{j,k}(t) \tag{2.34}$$

With the inverse wavelet transform, the nonstationary wind turbulence can be simulated. A good example to produce the nonstationary wind turbulence based on wavelet method can be seen in [122].

2.3.4 Nonstationary turbulence based on NONAR method

ARMA model and AR model

The autoregressive moving average (ARMA) model describes a stochastic process in terms of polynomials. It contains two parts, namely the autoregressive (AR) model and the moving average (MA) model. Compared to the wavelet method, ARMA-based methods are more easier to be implemented. A general description of ARMA model can be seen in [94] [183] and [201], which can be described as follows

$$a_0x_t + a_1x_{t-1} + a_2x_{t-2} + \cdots a_px_{t-p} = b_0\varepsilon_t + b_1\varepsilon_{t-1} + \cdots b_q\varepsilon_{t-q} \tag{2.35}$$

i.e.

$$\sum_{i=0}^{p} a_ix_{t-i} = \sum_{i=0}^{q} b_i\varepsilon_{t-i} \tag{2.36}$$

where x_t is a stationary time series, the subscript $t, t-1, \cdots, t-p$ means the time. $a_0, a_1 \cdots a_p$ and $b_0, b_1 \cdots b_q$ are the autoregressive coefficients and moving average coefficients, respectively. ε_t is a stationary white Gaussian noise with zero mean and finite variance σ_ε^2 .

Without loss of generality, assume that $a_0 = 1$, Equation (2.36) can be rewritten as

$$x_t + \sum_{i=1}^{p} a_ix_{t-i} = \sum_{i=0}^{q} b_i\varepsilon_{t-i} \tag{2.37}$$

To determine the autoregressive coefficients a_i, post multiplying Equation (2.37) by $x_{t-k}, (k > 0)$ and taking the expectation, one obtains

$$E(x_tx_{t-k})+a_1E(x_{t-1}x_{t-k})+\cdots a_pE(x_{t-q}x_{t-k}) = b_0E(\varepsilon_tx_{t-k})+b_1E(\varepsilon_{t-1}x_{t-k})+\cdots+b_qE(\varepsilon_{t-q}x_{t-k}) \tag{2.38}$$

Assume the mean value of x_t is zero and note that $E(\varepsilon_tx_{t-k}) = 0, k > 0$, Equation (2.38) leads to

$$R_{xx}(k) + a_1R_{xx}(k-1) + a_2R_{xx}(k-2) + \cdots a_pR_{xx}(k-p) = \begin{cases} 0 & , \quad k > q \\ b_q\sigma_\varepsilon^2 & , \quad k \leq q \end{cases} \tag{2.39}$$

where R_{xx} refers to the autocorrelation function of $x(t)$.

According to Equation (2.39), taking $k = q+1, q+2, \cdots q+p$, the following equations will be obtained

$$\begin{cases} R_{xx}(q+1) + a_1R_{xx}(q) + a_2R_{xx}(q-1) + \cdots a_pR_{xx}(q+1-p) = 0 \\ R_{xx}(q+2) + a_1R_{xx}(q+1) + a_2R_{xx}(q) + \cdots a_pR_{xx}(q+2-p) = 0 \\ \cdots \\ R_{xx}(q+p) + a_1R_{xx}(q+p-1) + a_2R_{xx}(q+p-2) + \cdots a_pR_{xx}(q) = 0 \end{cases} \tag{2.40}$$

which can be written in a matrix form by

$$\begin{bmatrix} R_{xx}(q+1) \\ R_{xx}(q+2) \\ \vdots \\ R_{xx}(q+p) \end{bmatrix} + \begin{bmatrix} R_{xx}(q) & R_{xx}(q-1) & \cdots & R_{xx}(q+1-p) \\ R_{xx}(q+1) & R_{xx}(q) & \cdots & R_{xx}(q+2-p) \\ \vdots & \vdots & \vdots & \vdots \\ R_{xx}(q+p-1) & R_{xx}(q+p-2) & \cdots & R_{xx}(q) \end{bmatrix} \begin{bmatrix} a_1 \\ a_2 \\ \vdots \\ a_p \end{bmatrix} = 0 \tag{2.41}$$

When the autocorrelation function R_{xx} is known, the autoregressive coefficients $a_1,\ a_2,\ \cdots,\ a_p$ can be determined by the above equation.

Next, to obtain the moving-average coefficients $b_0,\ b_1, \cdots b_q$, let

$$y_t = x_t + a_1 x_{t-1} + a_2 x_{t-2} + \cdots + a_p x_{t-p} \tag{2.42}$$

Post multiplying y_t with y_{t+k} and taking expectations, one obtains

$$\begin{aligned} R_{yy} &= E\left[(x_t + a_1 x_{t-1} + a_2 x_{t-2} + \cdots + a_p x_{t-p})(x_{t+k} + a_1 x_{t+k-1} + a_2 x_{t+k-2} + \cdots + a_p x_{t+k-p})\right] \\ &= E\left[(\sum_{i=0}^{p} a_i x_{t-i})(\sum_{j=0}^{p} a_j x_{t+k-j})\right] = \sum_{i=0}^{p}\sum_{j=0}^{p} a_i a_j R_{xx}(j-i) \end{aligned} \tag{2.43}$$

In a same way, let

$$z_t = b_0 \varepsilon_t + b_1 \varepsilon_{t-1} + b_2 \varepsilon_{t-2} + \cdots b_q \varepsilon_{t-q} \tag{2.44}$$

One obtains

$$\begin{aligned} R_{zz} &= E\left[(b_0 \varepsilon_t + b_1 \varepsilon_{t-1} + b_2 \varepsilon_{t-2} + \cdots b_q \varepsilon_{t-q})(b_0 \varepsilon_{t+k} + b_1 \varepsilon_{t+k-1} + b_2 \varepsilon_{t+k-2} + \cdots b_q \varepsilon_{t+k-q})\right] \\ &= \begin{cases} 0 & , \quad k > q \\ \sigma_\varepsilon^2 (b_0 b_k + b_1 b_{k+1} + \cdots + b_q b_{k+q}) \ , & \quad k \le q \end{cases} \end{aligned} \tag{2.45}$$

Note that $y_t = z_t$, i.e. $R_{yy} = R_{zz}$, Equation (2.43) and Equation (2.45) lead to

$$\sum_{i=0}^{p}\sum_{j=0}^{p} a_i a_j R_{xx}(j-i) = \begin{cases} 0 & , \quad k > q \\ \sigma_\varepsilon^2 (b_0 b_k + b_1 b_{k+1} + \cdots + b_q b_{k+q}) \ , & \quad k \le q \end{cases} \tag{2.46}$$

Based on Equation (2.46), the moving average coefficients can be obtained.

Assume that $q = 0$, ARMA model will be simplified to an AR model, which can be expressed as follows

$$x(t) + a_1 x_{t-1} + a_2 x_{t-2} + \cdots + a_p x_{t-p} = b_0 \varepsilon_t \tag{2.47}$$

Consequently, the autoregressive coefficients and moving average coefficients can be determined by

$$\begin{bmatrix} a_1 \\ a_2 \\ \vdots \\ a_p \end{bmatrix} = - \begin{bmatrix} 1 & R_{xx}(-1) & \cdots & R_{xx}(1-p) \\ R_{xx}(1) & 1 & \cdots & R_{xx}(2-p) \\ \vdots & \vdots & \vdots & \vdots \\ R_{xx}(p-1) & R_{xx}(p-2) & \cdots & 1 \end{bmatrix}^{-1} \begin{bmatrix} R_{xx}(1) \\ R_{xx}(2) \\ \vdots \\ R_{xx}(p) \end{bmatrix} \tag{2.48}$$

$$b_0 = \frac{1}{\sigma_\varepsilon} \sqrt{\sum_{i=0}^{p} \sum_{j=0}^{p} a_i a_j R_{xx}(j-i)} \tag{2.49}$$

As can be seen, the autoregressive parameters in the AR model depend on the autocorrelation function. For a stationary stochastic process, the autocorrelation function changes only with the lag k and it has no relations with time t. When a nonstationary stochastic process $X(t)$ is considered, the autocorrelation function will change with both lag k and time t. Therefore, the regressive parameters will be time-dependent.

NONAR method based on AR model

To simulate the nonstationary wind turbulence, a nonstationary autoregressive (NONAR) [67] method which is based on the AR model has been utilized in this work.

The NONAR model contains time dependent coefficients $A_i(t)$, $B(t)$ and can be expressed as follows:

$$\sum_{i=0}^{p} A_i(t) X_{t-i} = A_0(t) X_t + A_1(t) X_{t-1} + \cdots + A_p(t) X_{t-p} = B(t) \varepsilon_t \tag{2.50}$$

where X is the target time series (here, i.e. the nonstationary wind turbulence).

For every instant of time $\hat{t}$, in order to get the coefficients $A_i(\hat{t})$ and $B(\hat{t})$, Equation (2.50) is post multiplied by $X_{\hat{t}-j}$, $j = 1, 2, \cdots p$. After taking the expectation, one obtains

$$\sum_{i=0}^{p} A_i R_{XX}(\hat{t}-i, \hat{t}-j) = B R_{\varepsilon X}(\hat{t}, \hat{t}-j) \tag{2.51}$$

For simplicity, here and in the following, $A_i(\hat{t})$ and $B(\hat{t})$ are abbreviated as A_i and B, respectively. ε_t is assumed as a normalized Gaussian white noise process, i.e. $\varepsilon_t \sim N(0, 1)$.

$R_{\varepsilon X}$ is the cross-correlation function for X and ε_t, R_{XX} refers to the autocorrelation function of the signal X, which can be obtained by the following equation [158]

$$R_{XX}(\hat{t}_1, \hat{t}_2) = \int_{-\infty}^{+\infty} a(z, \hat{t}_1) a^*(z, \hat{t}_2) e^{i2\pi f(\hat{t}_1 - \hat{t}_2)} \Theta_u(z, f) \mathrm{d}f \tag{2.52}$$

where $a(z, \hat{t}_1)$, $a^*(z, \hat{t}_2)$ are the modulating functions at the instant of time $\hat{t}_1$, $\hat{t}_2$ and $\Theta(z, f)$ is the power spectral density, cf. Section 2.3.2.

Note that

$$R_{\varepsilon X}(\hat{t}, \hat{t} - j) = 0\,, \quad j = 1, 2, \cdots p \tag{2.53}$$

and assume $A_0(t)$ is equal to 1 for every instant of time, then the following equation will be obtained

$$\sum_{i=1}^{p} A_i R_{XX}(\hat{t} - i, \hat{t} - j) = -R_{XX}(\hat{t}, \hat{t} - j) \tag{2.54}$$

Rewriting Equation (2.54) in a matrix form, we obtain $\boldsymbol{A}$=$\boldsymbol{N}^{-1}\boldsymbol{M}$, where

$$\boldsymbol{A} = [A_1, A_2, \cdots, A_p]^T, \boldsymbol{M} = \left[R_{XX}(\hat{t}, \hat{t} - 1), R_{XX}(\hat{t}, \hat{t} - 2), \cdots, R_{XX}(\hat{t}, \hat{t} - p)\right]^T$$

$$\boldsymbol{N} = \begin{bmatrix} R_{XX}(\hat{t} - 1, \hat{t} - 1) & R_{XX}(\hat{t} - 2, \hat{t} - 1) & \cdots & R_{XX}(\hat{t} - p, \hat{t} - 1) \\ R_{XX}(\hat{t} - 1, \hat{t} - 2) & R_{XX}(\hat{t} - 2, \hat{t} - 2) & \cdots & R_{XX}(\hat{t} - p, \hat{t} - 2) \\ \vdots & \vdots & \vdots & \vdots \\ R_{XX}(\hat{t} - 1, \hat{t} - p) & R_{XX}(\hat{t} - 2, \hat{t} - p) & \cdots & R_{XX}(\hat{t} - p, \hat{t} - p) \end{bmatrix}$$

A similar analysis can be applied to obtain the coefficient $B = B(\hat{t})$, leading to

$$B = \sqrt{\sum_{i=0}^{p} A_i R_{XX}(\hat{t} - i, \hat{t})} \tag{2.55}$$

When the coefficients $A_i(\hat{t})$ and $B(\hat{t})$ for every instant of time $\hat{t}$ are obtained, the nonstationary wind turbulence can be simulated according to Equation (2.50). An example of the nonstationary wind turbulence produced by the NONAR method can be seen in Figure 2.7, in which the average wind speed $u_0 = 16$ m/s.

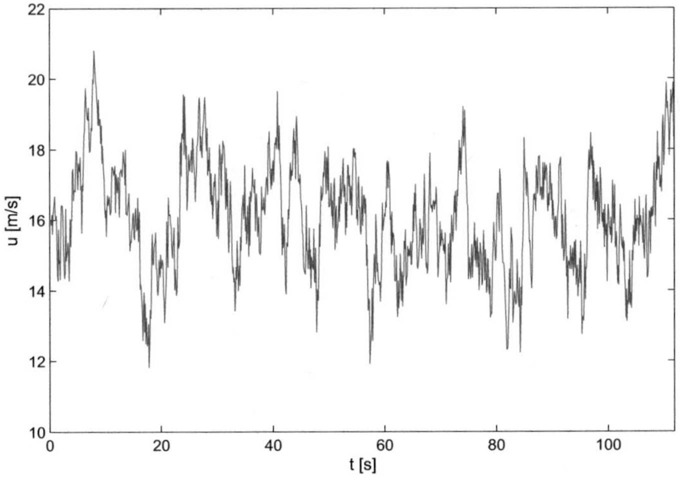

Figure 2.7: Wind with nonstationary turbulence at $u_0 = 16$ m/s

2.4 Gust scenario

A gust is considered as a maximal deviation of the turbulent wind process between two zero crossings. It is usually utilized to model certain extreme wind conditions.

2.4.1 Gust shapes

So far, a lot of different gust shapes have been proposed in the literature (e.g. the 'Mexican hat' model [30], the '1-cos' [193] [127] model, the 'Chinese hat' [84] and the 'rugby-ball' [54]). They are obtained either by convention or by selecting a reference gust out of certain statistical distributions. However, due to the nonlinear character of the relationship between the wind speed and vehicle dynamics, the simplification introduced by assuming a deterministic gust seems difficult to quantify [161].

Figure 2.8 shows an example of the famous 'Mexican hat' gust model, which has been used to calibrate the gust magnitude in the turbine design and can be described by the following wavelet function [128] [30]:

$$\varphi(t) = \frac{1}{\sqrt{\Gamma(5/2)}}(1 - t^2)e^{(-t^2/2)} \tag{2.56}$$

The sudden change in 'Mexican hat' gust model may cause extreme change in loads. To characterize this type of gust, several parameters, such as the gust amplitude, the rise time and the lapse time have to be determined, see [30] [139].

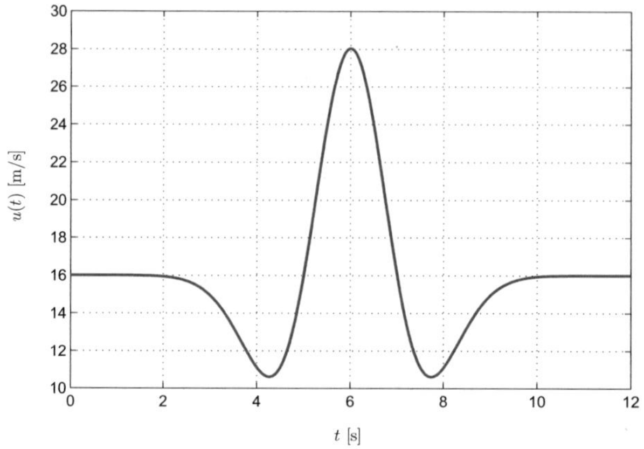

Figure 2.8: An example of the 'Mexican hat' gust model

Another simple gust model, namely the '1-cos' model, which has been usually utilized in aircraft engineering [129] [193], can be expressed as follows

$$u(t) = \begin{cases} 0 & , \quad t < t_B - \frac{T}{2}\ ,\ t > t_B - \frac{T}{2} \\ \frac{A}{2}\left[1 - \cos\left(\pi\left(1 + \frac{t-t_B}{T}\right)\right)\right] & , \quad t_B - \frac{T}{2} \ < \ t < t_B + \frac{T}{2} \end{cases} \tag{2.57}$$

where A is the gust amplitude and t_B refers to the time at the maximal gust speed.

The variance σ_T of T can be obtained by the following equation [127]:

$$3\sigma_T = 0.5T \tag{2.58}$$

Besides, according to [34], the gust factor g_f, which is defined as the ratio between the peak value and mean value of the wind speed, can be set to 1.8.

In fact, the estimation of gust amplitude as well as other characteristic parameters of the gust is not an easy work due to the great variability of wind speed [31]. A standard but simple way to obtain the gust amplitude in weather predictions is to consider a constant ratio between the maximal gust and the mean wind speed [31]. In [34], an empirical gust model, which considers the gust as a constant mean wind speed superimposed by a '1-cos' gust amplitude has been used. The gust amplitude and duration are considered as deterministic values for the risk analysis. More examples of the deterministic gust models can be seen in [22] [193] and [127]. The underlying assumptions for these traditional gust models are based on a specific gust process, namely based on a deterministic gust shape.

In reality, the gust is always a stochastic process. To simulate more realistic gust scenarios, a physically based stochastic gust model has to be implemented. Recently, an efficient way to simulate the stochastic gust model by means of the so called constrained-simulation approach has been proposed for wind turbine reliability analysis, cf. [26]. The main advantage of this method is that the turbulent wind process with the superimposed gust characteristic is statistically indistinguishable from the natural wind process. Based on this approach, the total gust speed can be defined as follows [203]:

$$u(t) = (u_0 + u'(t)) + u_B(t - t_B)(A - u'(t_B)) - \frac{\dot{u}_B(t - t_B)\dot{u}'(t_B)}{\ddot{u}_B(t_B)} \tag{2.59}$$

where $u_B(t)$ is the gust characteristic based on the coherence function. It allows a simulation of gusts with different durations for the moving vehicle. As can be seen, the first term of Equation (2.59) contains the nonstationary wind turbulence and the second term generates a mean gust shape with a target amplitude A. Additionally, the third term makes sure that the gust reaches at a maximum at time t_G. According to [161], the gust characteristic can be given by

$$u_B = \exp\left(-\frac{1}{2Tu_0}\sqrt{(C_x\Delta x)^2 + (C_y\Delta y)^2}\right) \tag{2.60}$$

in which C_x and C_y are the corresponding coherence coefficients, the vertical direction is not considered. Δx and Δy are the relative distances of two points in space, which need to be determined by the following step.

Considering a vehicle running under a turbulent crosswind, to determine Δx and Δy, the famous Taylor's 'frozen turbulence' hypothesis has to be applied [59]. It assumes that the turbulence field can be considered as frozen in both space and time and the field of turbulence past a fixed point can be taken entirely by the mean flow.

Figure 2.9 shows the geometrical relationship of a vehicle running in the turbulent wind. Assume that the vehicle moves along a straight line with a constant vehicle speed v_0 and the wind speed at a fixed point is described by a fixed Cartesian coordinate with three directions (x, y, z). P is a point fixed in the vehicle. With a time delay Δt, the physical point P will move to P'. P'_e is an equivalent point for P' in the frozen turbulent field. So the coherence function of P' and P for a moving vehicle can be obtained by calculating the coherence function between the equivalent point P'_e and P.

Therefore, one obtains

$$\Delta x = (u_0 + v_0 \sin(\alpha_\omega))\Delta t, \quad \Delta y = (v_0 \cos(\alpha_\omega))\Delta t \tag{2.61}$$

and Equation (2.60) leads to

$$u_B = \exp\left(-\frac{1}{2Tu_0}\sqrt{(C_x(u_0 + v_0 \sin(\alpha_\omega))\Delta t)^2 + (C_y(v_0 \cos(\alpha_\omega))\Delta t)^2}\right) \tag{2.62}$$

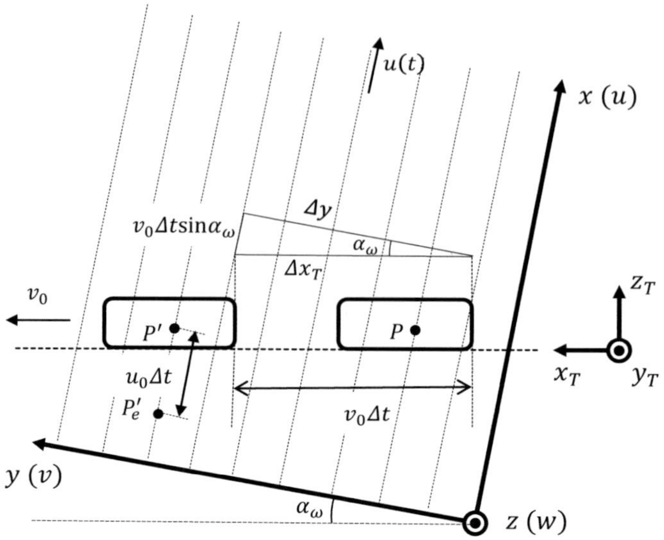

Figure 2.9: Geometry analysis for a vehicle moving through the turbulent wind

when a varying vehicle speed is considered, the coherence function has to be obtained by an integral

$$u_B^* = \exp\left(-\frac{1}{2Tu_0}\sqrt{(C_x(u_0\Delta t + \int_0^{\Delta t} v(\tau)\sin(\alpha_\omega)\mathrm{d}\tau))^2 + (C_y\int_0^{\Delta t} v(\tau)\cos(\alpha_\omega)\mathrm{d}\tau)^2}\right) \tag{2.63}$$

For a special case, i.e. the vehicle accelerates or decelerates with a constant acceleration a_c, u_B^* is given by

$$u_B^* = \exp\left(-\frac{1}{2Tu_0}\sqrt{C_x^2\Big[(v_0\Delta t + \frac{1}{2}a_c\Delta t^2)\sin(\alpha_\omega) + u_0\Delta t\Big]^2 + C_y^2\Big[(v_0\Delta t + \frac{1}{2}a_c\Delta t^2)\cos(\alpha_\omega)\Big]^2}\right) \tag{2.64}$$

When u_B^* is obtained, the gust model can be well simulated based on Equation (2.59). A typical realization of the gust with nonstationary wind turbulence at $u_0 = 14$ m/s, $v_0 =$ 160 km/h based on the above method can be seen in Figure 2.10.

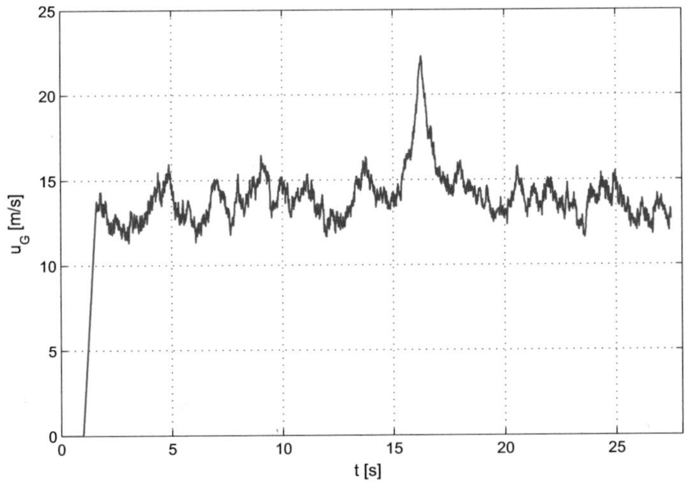

Figure 2.10: A gust model with nonstationary turbulence

2.4.2 Probabilistic characteristics of the gust

Following the statistical method given in [27] for a peak-over-threshold process, the occurrence for a gust amplitude above an absolute threshold $|A_0|$ can be expressed as follows:

$$\mathrm{Z}^+(A_0) = -\ddot{u}(t)\delta(\dot{u}(t))e(u(t) - A_0)\ , \ \ u(t) > 0 \tag{2.65}$$

and

$$\mathrm{Z}^-(A_0) = \ddot{u}(t)\delta(\dot{u}(t))e(-u(t) - A_0)\ , \ \ u(t) < 0 \tag{2.66}$$

where $e(t)$ is a unit step function at $t = 0$, $\delta(t)$ refers to a delta function. $\mathrm{Z}^+(A_0)$ and $\mathrm{Z}^-(A_0)$ are the occurrence for the gust amplitude beyond the threshold value A_0 and $-A_0$, respectively.

The mean number of the occurrence for $u(t) > A_0$ in the time interval $[0, T_0]$ can be expressed as

$$N^+(A_0) = E(N^+_{T_0}(A_0)) = E(\int_0^{T_0} \mathrm{Z}^+(A_0)\mathrm{d}t) = \int_0^{T_0} E(\mathrm{Z}^+(A_0))\mathrm{d}t \tag{2.67}$$

where

$$\begin{aligned} E(\mathrm{Z}^+(A_0)) &= \int\int\int -\ddot{u}(t)\delta(\dot{u}(t))e(u(t) - A_0)\, f(u, \dot{u}, \ddot{u}, t)\mathrm{d}u\mathrm{d}\dot{u}\mathrm{d}\ddot{u} \\ &= -\int_{A_0}^{+\infty}\int_{-\infty}^{+\infty} \ddot{u}\, f(u, 0, \ddot{u}, t)\mathrm{d}u\mathrm{d}\ddot{u} \end{aligned} \tag{2.68}$$

$f(u, \dot{u}, \ddot{u})$ is the joint density function of $u(t), \dot{u}(t)$ and $\ddot{u}(t)$.

Similarly, for the event of $u(t) < -A_0$

$$N^-(A_0) = E(N_{T_0}^-(A_0)) = E(\int_0^{T_0} Z^-(A_0)\mathrm{d}t) = \int_0^{T_0} E(Z^-(A_0))\mathrm{d}t \tag{2.69}$$

$$\begin{aligned} E(Z^-(A_0)) &= \int\int\int \ddot{u}(t)\delta(\dot{u}(t))e(-u(t) - A_0)\, f(u, \dot{u}, \ddot{u})\mathrm{d}u\mathrm{d}\dot{u}\mathrm{d}\ddot{u} \\ &= \int_{-\infty}^{-A_0}\int_{-\infty}^{+\infty} \ddot{u}\, f(u, 0, \ddot{u})\mathrm{d}u\mathrm{d}\ddot{u} \end{aligned} \tag{2.70}$$

The distribution of the event that the gust amplitude $|A| \leq |A_0|$ can be deduced as follows

$$\begin{aligned} F(A_0) &= P(A \leq A_0) = 1 - P(A > A_0) = 1 - \frac{N(A_0)}{N(0)} = 1 - \frac{N^+(A_0) + N^{-1}(A_0)}{N(0)} \\ &= 1 - \frac{1}{N(0)}\int_0^{T_0}\int_{A_0}^{+\infty}\int_{-\infty}^{+\infty} \ddot{u}f(u, 0, \ddot{u})\mathrm{d}u\mathrm{d}\ddot{u}\mathrm{d}t + \int_0^{T_0}\int_{-\infty}^{-A_0}\int_{-\infty}^{+\infty} \ddot{u}\, f(u, 0, \ddot{u})\mathrm{d}u\mathrm{d}\ddot{u}\mathrm{d}t \end{aligned} \tag{2.71}$$

where $N(0)$ is the total occurrence at the threshold value of zero, which can be described as follows

$$N(0) = N^+(0) + N^-(0) \tag{2.72}$$

The probability density function of $F(A_0)$ can be obtained by

$$f(A_0) = \frac{dF(A_0)}{dA_0} \tag{2.73}$$

In principle $f(A_0)$ can be evaluated when the joint density function $f(u, \dot{u}, \ddot{u})$ is obtained, cf. Equation (2.71) and (2.73). In practice it will be difficult, since the density function $f(u, \dot{u}, \ddot{u})$ is not so easy to obtain [27]. One feasible way to evaluate the density function is under the assumption of a stationary process (e.g. Gaussian process), cf. [27] [35] [168] and [124]. It would be also possible to evaluate $f(A_0)$ on the basis of measured wind data, cf. [24] [66].

Recently, the analysis for the probability density function of extreme values in nonstationary cases are found in [108] [23] and [78]. They consider the distributions of extreme values as a function of time, which is based on a generalized extreme value (GEV) distribution with time-independent parameters. However, these methods are essentially based on measured wind data for a chosen time interval. The PDF of wind data in each time interval is still analyzed under the assumption of a stationary process and the fitting results (i.e. the PDF with time-dependent coefficients) show only the trend of PDF with realistic time (e.g. each year or each month).

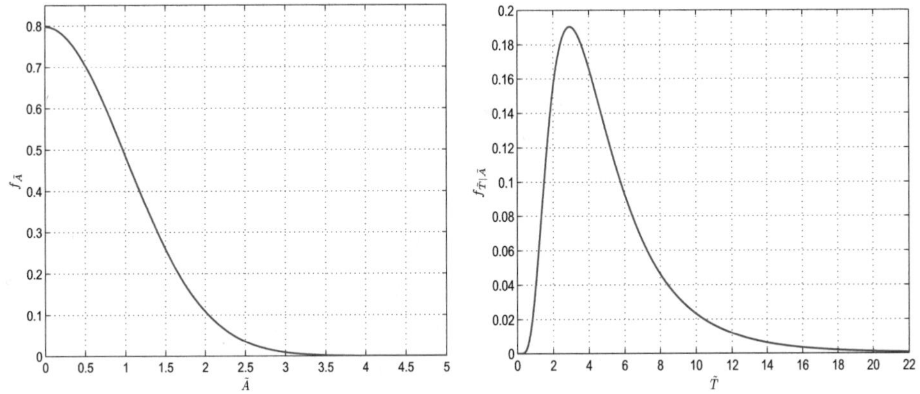

(a) Probability density function for $\tilde{A}$ (b) Conditional probability density function for $\tilde{T}$

Figure 2.11: Probability density functions for normalized gust amplitude $\tilde{A}$ and gust duration $\tilde{T}$

Consider that the risk analysis in this work is usually carried out based on a single gust event. For each gust scenario, the failure probabilities of vehicles under different conditions (e.g. wet/dry road, different wind angles) are expected as functions of different wind and vehicle speeds, but rather than the time. Therefore, the PDFs of gust parameters (A and T) for the mean gust shape are thought as time-independent and the gust parameters on the whole are considered as a sequence of independent and identically distributed random variables. The total gust shapes are simulated as mean gust shapes combined with the nonstationary turbulence, cf. Equation (2.59). In addition, based on measurements [66], it has been also found that the distribution of normalized gust amplitudes for a series of measured wind data can be fitted as a half-Gaussian distribution, of which the probability density function can be defined by [203]

$$f_{\tilde{A}} = \sqrt{2/\pi} \exp(-\tilde{A}^2/2) \tag{2.74}$$

where $\tilde{A}$ is a normalized gust amplitude. The relationship between $\tilde{A}$ and the actual amplitude A can be described as follows

$$A = \tilde{A}\sigma_{u'} \tag{2.75}$$

where

$$\sigma_{u'} = \frac{u_0}{\ln(z/z_0)} \left(1.0 - 5 \times 10^{-5} \left(\log\left(\frac{z_0}{0.05}\right) + 2\right)^7\right) \tag{2.76}$$

For a given $\tilde{A}$, the conditional distribution $f_{\tilde{T}|\tilde{A}}$ of the normalized duration $\tilde{T}$ can be fitted by the log-normal probability function with a standard deviation $\sigma_{Ln(\tilde{T})} = 0.6$, cf. [66] [69] and [203].

$$f_{\tilde{T}|\tilde{A}} = \exp\left(-\frac{1}{2}\left(\frac{\ln(\tilde{T}) - \ln\left(0.95\tilde{A}^{1.42}\right)}{\sigma_{\ln(\tilde{T})}}\right)\right) \Big/ \left(\sqrt{2\pi}\tilde{T}\sigma_{\ln(\tilde{T})}\right) \tag{2.77}$$

in which

$$\tilde{T} = T \big/ \bar{T} \tag{2.78}$$

$\bar{T}$ is the mean gust duration and can be determined based on the occurrence rate by the following equation [24] [168]

$$\bar{T} = \frac{1}{2N^{+}(0)} \tag{2.79}$$

Figure 2.11 (a) illustrates the probability density function of the normalized gust amplitude $\tilde{A}$ based on half-Gaussian distribution and Figure 2.11 (b) shows an example for the probability density function of $\tilde{T}$ at $\tilde{A} = 2.84$ [69].

It should be noted that the distribution described above are usually based on the analysis of wind data measured at certain sites. Although it is applicable for most places, it still needs to be verified in the practical use (i.e. for a specific location). In this regard, other distribution models might be considered, e.g. the generalized extreme value distribution like Type I and III [100] [153].

3 Wind excitation analysis

In this chapter, wind excitations caused by the nonstationary crosswind are studied. Considering the fact that the realistic wind is distributed over the vehicle body, a vehicle with spatial dimension under the excitation of turbulent wind is analyzed. For a general case of the vehicle running with a varying vehicle speed under the excitation of strong crosswinds with turbulence, the equivalent spectrum which contains the aerodynamic admittance function is utilized to compute the wind forces. In addition, wind excitations caused by vortex and buffeting phenomena are studied. At last, the emerged Hilbert-Huang transform (HHT) method is applied to analyze the nonstationary wind excitation.

3.1 Crosswind excitation

The investigations on aerodynamic forces of vehicles under strong crosswinds are usually performed by means of wind tunnel tests [38] [40], full scale tests [142] [15] and numerical methods (e.g. CFD simulations) [73] [50]. In the past decades, many wind tunnel experiments were carried out to obtain the aerodynamic forces and pressure distributions. Most of them are performed with still vehicle models. It is considered that if the pressure on the vehicle body can be measured and integrated over the vehicle surface, the aerodynamic force caused by the crosswind can be estimated [113]. However, measuring of the pressure over the whole vehicle surface is a laborious task. On the other hand, the full scale tests are complex to carry out since they rely on the conditions of the uncontrolled nature of the realistic wind and need very careful handing to obtain meaningful results [15]. Therefore, numerical methods together with experimental verifications are more and more widely used. A comparison of the above three different methods to assess the wind induced forces can be seen in [189].

3.1.1 Relative wind speed

Before studying the wind force caused by strong crosswinds, the relative wind speed is analyzed. Figure 3.1 represents a vector diagram of the relative wind speed for a moving vehicle. In Figure 3.1 (a), the mean wind speed and vehicle speed are considered, corresponding relationship of the variables can be expressed as

$$V_{res}(t)^2 = \overline{V}^2_{res}(t) = v_0^2 + u_0^2 + 2u_0v_0\sin(\alpha_\omega) \tag{3.1}$$

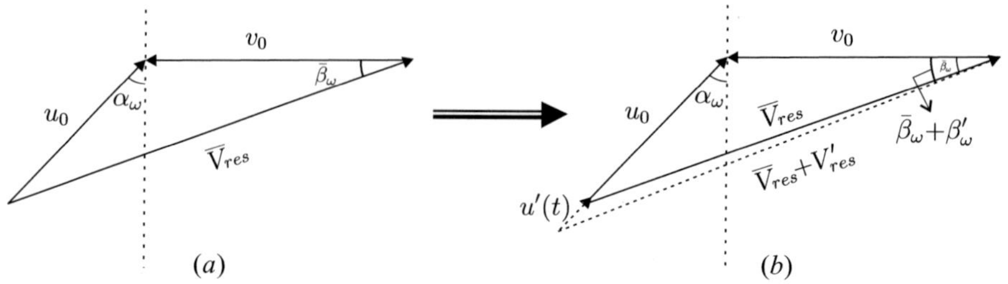

Figure 3.1: Relative wind speed V_{res} for the moving vehicle w. or w.o. nonstationary turbulence

where $V_{res}(t)$ refers to the relative wind speed, $\overline{V}_{res}$ is the mean value of $V_{res}(t)$, α_ω is the wind angle, v_0 and u_0 are the mean vehicle speed and wind speed, respectively.

When the turbulent wind is taken into consideration, the relative wind speed and angle will change due to the wind turbulence as can be seen in Figure 3.1(b)

$$u(t) = u_0 + u'(t) \tag{3.2}$$

$$V_{res}(t) = \bar{V}_{res} + V'_{res} \tag{3.3}$$

$$\begin{aligned} V_{res}^2(t) &= (u(t)\sin\alpha_\omega + v_0)^2 + (u(t)\cos\alpha)^2 \\ &= ((u_0 + u'(t))\sin\alpha_\omega + v_0)^2 + ((u_0 + u'(t))\cos\alpha)^2 \\ &= u_0^2 + 2u_0u'(t) + u'(t)^2 + v_0^2 + 2u_0v_0\sin\alpha_\omega + 2u'(t)v_0\sin\alpha_\omega \end{aligned} \tag{3.4}$$

$$\begin{aligned} \beta_\omega(t) &= \bar{\beta}_\omega + {\beta'}_\omega = \arctan\left(\frac{u(t)\cos\alpha_\omega}{v_0 + u(t)\sin\alpha_\omega}\right) \\ &= \arctan\left(\frac{u_0\cos\alpha_\omega + u'(t)\cos\alpha_\omega}{v_0 + u_0\sin\alpha_\omega + u'(t)\sin\alpha_\omega}\right) \end{aligned} \tag{3.5}$$

where $u(t)$ is the total wind speed, $u'(t)$ is the wind turbulence, V'_{res} refer to the turbulent component of the relative wind speed, $\bar{\beta}_\omega$ and β'_ω are mean part and fluctuating part of the relative wind angle $\beta_\omega(t)$, respectively.

In reality, the vehicle speed always changes due to various conditions, e.g. accelerating or braking. In order to study the crosswind stability of vehicles running under the non-stationary crosswind in a more realistic scenario, a vehicle with varying speed has to be considered.

Figure 3.2 shows the corresponding vector diagram of the relative wind speed in the situation with both nonstationary wind turbulence and varying vehicle speed. It is assumed

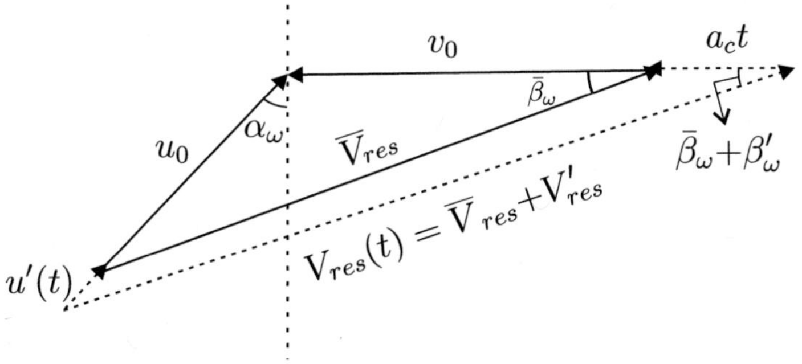

Figure 3.2: Vector diagram of the relative wind speed under nonstationary situation (including nonstationary wind turbulence and varying vehicle speed)

that the vehicle accelerates and decelerates with a constant acceleration. Corresponding relationships between the parameters can be described as follows

$$v(t) = v_0 + a_c t \tag{3.6}$$

$$\begin{aligned} V_{res}^2(t) &= v(t)^2 + u(t)^2 + 2u(t)v(t)\sin(\alpha_\omega) \\ &= (v_0 + a_c t)^2 + (u_0 + u'(t))^2 + 2(v_0 + a_c t)(u_0 + u'(t))\sin(\alpha_\omega) \\ &= v_0^2 + 2v_0 a_c t + a_c^2 t^2 + u_0^2 + 2u_0 u'(t) + u'(t)^2 + 2v_0 u_0 \sin(\alpha_\omega) \\ &\quad + 2v_0 u'(t)\sin(\alpha_\omega) + 2u_0 a_c t \sin(\alpha_\omega) + 2u'(t) a_c t \sin(\alpha_\omega) \end{aligned} \tag{3.7}$$

$$\beta_\omega(t) = \beta_\omega + \beta'_\omega = \arctan\left(\frac{u(t)\cos\alpha_\omega}{v_0 + a_c t + u(t)\sin\alpha_\omega}\right) \tag{3.8}$$

where a_c refers to the acceleration or deceleration of the vehicle speed.

3.1.2 Turbulent wind excitation and admittance function

In practice, it is conventional to write the wind force acting on a vehicle as a concentrated force. This implicitly assumes that the dimension of the vehicle is neglected, i.e. the vehicle can be regarded as a point-like object. According to [11], the concentrated force can be defined as follows:

$$F(t) = \frac{1}{2}\rho_L A_C C_F(\beta_\omega(t)) V_{res}^2(t) \tag{3.9}$$

where $F(t)$ refers to the wind force, ρ_L is the air density and A_C refers to the characteristic area of the vehicle. $C_F(\beta_\omega(t))$ is the non-dimensional aerodynamic coefficient which is usually a nonlinear function of the relative wind angle $\beta_\omega(t)$ and has to be obtained by CFD simulation or wind tunnel tests.

For a static vehicle under the excitation of crosswind at mean wind speed u_0, the mean wind force can be written as

$$F_0 = \frac{1}{2}\rho_L A_C C_F(\bar{\beta}_\omega) V_{res}^2(t) = \frac{1}{2}\rho_L A_C C_F u_0^2 \tag{3.10}$$

where C_F is the abbreviation of $C_F(\bar{\beta}_\omega)$.

When the wind turbulence is taken into account, the wind force in an ideal case can be defined according to the steady formulation [39], i.e.

$$F(t) = \frac{1}{2}\rho_L A_C C_F (u_0 + u'(t))^2 = \frac{1}{2}\rho_L A_C C_F \left(u_0^2 + 2u_0 u'(t) + u'^2(t)\right) \tag{3.11}$$

Considering a lag time τ for $F'(t)$ and $u'(t)$, Equation (3.11) can be written as

$$F(t+\tau) = \frac{1}{2}\rho_L A_C C_F \left(u_0^2 + 2u_0 u'(t+\tau) + u'^2(t+\tau)\right) \tag{3.12}$$

Omitting the second-order terms of Equations (3.11) and (3.12), and multiplying the two equations. After taking the expectation, the following relationship can be obtained

$$R_{FF}(\tau) = \frac{1}{4}\rho_L^2 A_C^2 C_F^2 \left(u_0^4 + 4u_0^2 R_{u'u'}(\tau)\right) \tag{3.13}$$

where $R_{FF}(\tau)$ and $R_{u'u'}(\tau)$ are the autocorrelation function of $F(t)$ and $u'(t)$.

Taking the Fourier transform of Equation (3.13), the relationship of PSD between the wind force and wind speed can be obtained as

$$S_{FF}(f) = \frac{1}{4}\rho_L^2 A_C^2 C_F^2 \left(u_0^4 \delta(f) + 4u_0^2 S_{u'u'}(f)\right) \tag{3.14}$$

where $S_{FF}(f)$ and $S_{u'u'}(f)$ are the PSD of wind force and turbulence, respectively.

In fact, for a vehicle with spatial dimension, a spatial average of the squared wind velocity has to be done over the whole characteristic area of the vehicle [203]. Therefore, Equation (3.11) has to be changed as:

$$\begin{aligned} F(t) &= \frac{1}{A_C}\int_{A_C} \frac{1}{2}\rho_L A_C C_F (u_0 + u'(t))^2 \mathrm{d}A_C \\ &= \frac{1}{2}\rho_L C_F \int_{A_C} \left(u_0^2 + 2u_0 u'(t) + u'^2(t)\right) \mathrm{d}A_C \\ &= \frac{1}{2}\rho_L A_C C_F \left(u_0^2 + 2\frac{u_0}{A_C}\int_{A_C} u'(x,y,z,t)\,\mathrm{d}A_C + \frac{1}{A_C}\int_{A_C} u'^2(x,y,z,t)\,\mathrm{d}A_C\right) \end{aligned} \tag{3.15}$$

Similarly, note that Equation (3.15) can be written with a later time τ

$$F(t+\tau) = \frac{1}{2}\rho_L A_C C_F \left(u_0^2 + 2\frac{u_0}{A_C}\int_{A_C} u'(x,y,z,t+\tau)\,\mathrm{d}A_C + \frac{1}{A_C}\int_{A_C} u'^2(x,y,z,t+\tau)\,\mathrm{d}A_C\right) \tag{3.16}$$

Neglecting the second-order terms of Equation (3.15) and (3.16) and multiplying them, one obtains

$$R_{FF}(\tau) = \frac{1}{4}\rho_L^2 A_C^2 C_F^2 \left(u_0^4 + \frac{4u_0^2}{A_c^2}\int_{A_C}\int_{A'_C} R_{u'u'}(x,y,z,x',y',z',\tau)\mathrm{d}A_C\mathrm{d}A'_C\right) \tag{3.17}$$

where

$$R_{FF}(\tau) = \overline{F(t)F(t+\tau)},\, R_{u'u'}(\tau) = \overline{u'(x,y,z,t)u'(x,y,z,t+\tau)} \tag{3.18}$$

Based on the Fourier transform, Equation (3.17) leads to

$$S_{FF} = \frac{1}{4}\rho_L^2 A_C^2 C_F^2 \left(u_0^4\,\delta(f) + \frac{4u_0^2}{A_C^2}\int_{A_C}\int_{A'_C} S_{u'u'}(x,y,z,x',y',z',f)\mathrm{d}A_C\mathrm{d}A'_C\right) \tag{3.19}$$

Note that the cross spectral density $S_{u'u'}(x,y,z,x',y',z',f)$ can be determined by the coherence function and the wind PSD, i.e.

$$S_{u'u'}(x,y,z,x',y',z',f) = \mathrm{Coh}(f)S_{u'u'}(f) \tag{3.20}$$

where $\mathrm{Coh}(f)$ is the coherence function of $u'(x,y,z)$ and $u'(x',y',z')$, cf. Equation (2.28).

Combining Equation (3.19) and (3.20), one obtains

$$\begin{aligned} S_{FF} &= \frac{1}{4}\rho_L^2 A_C^2 C_F^2 \left(u_0^4\,\delta(f) + \frac{4u_0^2}{A_C^2}\int_{A_C}\int_{A'_C} \mathrm{Coh}(f)S_{u'u'}(f)\mathrm{d}A_C\mathrm{d}A'_C\right) \\ &= \frac{1}{4}\rho_L^2 A_C^2 C_F^2 \left(u_0^4\delta(f) + 4u_0^2\chi_F^2(f)S_{u'u'}(f)\right) \end{aligned} \tag{3.21}$$

where

$$\chi_F^2(f) = \frac{1}{A_c^2}\int_{A_C}\int_{A'_C} \mathrm{Coh}(f)\mathrm{d}A_C\mathrm{d}A'_C \tag{3.22}$$

The item $\chi_F^2(f)$, which is also called the aerodynamic admittance function, has a significant meaning for the calculation of wind force. Comparing Equation (3.21) and (3.14), it can be found that the admittance function $\chi_F^2(f)$ accounts for the lack of spatial correlation of the wind turbulence on the vehicle surface. It represents a scaling factor for the simulation of the vehicle enveloped by turbulent wind with full spatial correlation.

For a moving vehicle with an approximate rectangular area under the turbulent wind, the aerodynamic admittance becomes [59]:

$$\chi_F^2(f) = \int_0^{H_T}\int_0^{L_T}\int_0^{H_T}\int_0^{L_T} \frac{1}{L_T^2 H_T^2} \cdot \mathrm{Coh}(f)\ \mathrm{d}z_T \mathrm{d}x_T\, \mathrm{d}z'_T \mathrm{d}x'_T \tag{3.23}$$

where L_T and H_T refer to the length and height of the vehicle, respectively. x_T and z_T are the variables in the vehicle body coordinate system, cf. Figure 2.9.

According to Bearman [20], the computation of the aerodynamic admittance can also be reduced to a double dimensional integral as

$$\chi_F^2(f) = \frac{4}{L_T^2 H_T^2} \int_0^{H_T}\int_0^{L_T} (L_T - x_T)(H_T - z_T)\,\mathrm{Coh}(f)\ \mathrm{d}z_T \mathrm{d}x_T \tag{3.24}$$

In practice, the admittance function can be obtained either by experimental tests [188] or by numerical models [11]. If the wind forces are calculated in the time domain, the corresponding aerodynamic weighting function has to be utilized. The relationship of the aerodynamic admittance $\chi_F^2(f)$ and weighting function $h_F(\tau)$ can be expressed by [188]

$$h_F(\tau) = \int_0^{+\infty} \chi_F^2(f) e^{2\pi i f} \mathrm{d}f \tag{3.25}$$

A good example of the wind force calculated in the time domain by using the weighting function can be seen in [14]. Due to the fact that the aerodynamic admittance functions for different railway vehicles usually appear to follow the same trend regardless of the yaw angle or train type, Sterling [188] examined the aerodynamic admittance and weighting function with a general curve given by Baker [11]. A series of experimental data from different wind-tunnel and full-scale tests on various vehicles have been studied and a relatively simple formula for the aerodynamic admittance/weighting function for different vehicles has been established.

Figure 3.3 illustrates the aerodynamic admittances of the side force and lift force for a certain train (Class 365) under the condition that the yaw angle is equal to 90°. The corresponding aerodynamic weighting functions are shown as well.

3.1.3 Nonstationary wind excitations acting on a moving vehicle

Not spatially extended vehicles with constant vehicle speed

Following Cooper [59] and Baker [11] [14], it is assumed that the fluctuations of wind

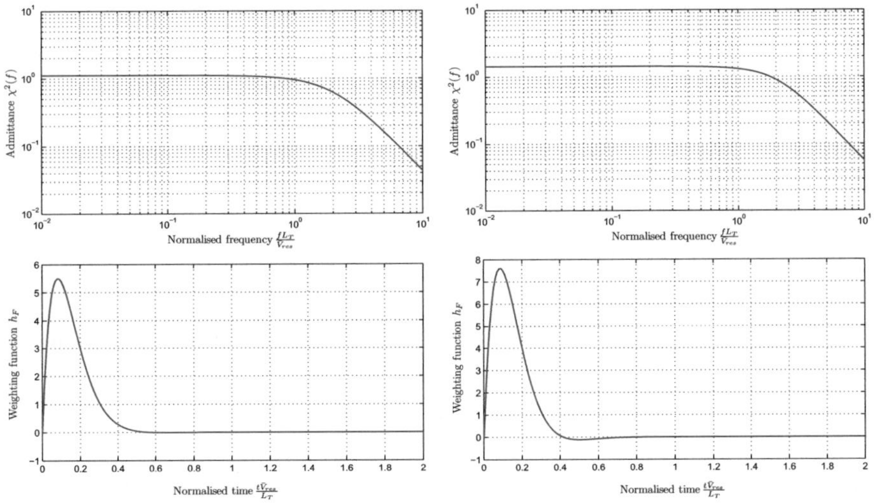

(a) Side force (b) Lift force

Figure 3.3: Aerodynamic admittance and weighting function of the side force and lift force for the train (Class 365) [188]

forces are caused by wind turbulence. Without consideration of the spatial extension of the vehicle, the concentrated force based on Equation (3.9) can be rewritten as

$$F(t) = \bar{F} + F' = \frac{1}{2}\rho_L A_C C_F(\beta_\omega(t))(\bar{V}_{res} + V'_{res})^2 \tag{3.26}$$

where $\bar{F}$ and F' are the mean part and fluctuating part of $F(t)$, respectively.

Based on the Taylor series expansion, $C_F(\beta_\omega(t))$ can be expressed by

$$C_F(\beta_\omega(t)) \approx C_F(\bar{\beta}_\omega) + \frac{\mathrm{d}C_F}{\mathrm{d}\beta_\omega}\beta'_\omega = C_F + \frac{\mathrm{d}C_F}{\mathrm{d}\beta_\omega}\beta'_\omega \tag{3.27}$$

and thus Equation (3.26) leads to

$$F(t) = \bar{F} + F' = \frac{1}{2}\rho_L A_C(C_F + \frac{\mathrm{d}C_F}{\mathrm{d}\beta_\omega}\beta'_\omega)(\bar{V}_{res} + V'_{res})^2 \tag{3.28}$$

Assume that the wind direction is perpendicular to the vehicle's running direction and the vehicle speed is a constant. From Figure 3.1(b), the following equations can be obtained

$$v_0{}^2 + u_0{}^2 = \bar{V}_{res}^2 \tag{3.29}$$

$$v_0{}^2 + (u_0 + u')^2 = (\bar{V}_{res} + V'_{res})^2 \tag{3.30}$$

$$\tan \beta_\omega = \frac{u_0}{v_0} \tag{3.31}$$

$$\tan(\beta_\omega + \beta'_\omega) = \frac{u_0 + u'}{v_0} \tag{3.32}$$

Based on Equation (3.28)$\sim$ (3.32) and neglecting the second-order item, the following results are deduced

$$\begin{aligned} \bar{F} + F' &= \frac{1}{2}\rho_L A_C (C_F \bar{V}_{res}^2 + C_F \cdot 2\bar{V}_{res}V'_{res} + \frac{\mathrm{d}C_F}{\mathrm{d}\beta_\omega}\beta'_\omega \cdot \bar{V}_{res}^2 + \frac{\mathrm{d}C_F}{\mathrm{d}\beta_\omega}\beta'_\omega \cdot 2\bar{V}_{res}V'_{res}) \\ &\approx \underbrace{\frac{1}{2}\rho_L A_C C_F \bar{V}_{res}^2}_{\text{Mean Force}} + \underbrace{\rho_L A_C C_F u_0 u' (1 + \frac{1}{2C_F}\frac{\mathrm{d}C_F}{\mathrm{d}\beta_\omega}\cot\beta_\omega)}_{\text{Turbulence Force}} \end{aligned} \tag{3.33}$$

By equating the mean and fluctuating parts, one obtains

$$F' = \rho_L A_C C_F u_0 (1 + \frac{1}{2C_F}\frac{\mathrm{d}C_F}{\mathrm{d}\beta_\omega}\cot\beta_\omega)u' \tag{3.34}$$

Thus, the PSD of fluctuating wind force acting on a moving vehicle can be obtained

$$S_{F'F'}(f) = (\rho_L A_C C_F u_0 (1 + \frac{1}{2C_F}\frac{\mathrm{d}C_F}{\mathrm{d}\beta_\omega}\cot\beta_\omega))^2 S_{u'u'}(f) \tag{3.35}$$

Spatially extended vehicles with constant vehicle speed

Considering the realistic effect of the wind turbulence acting on the vehicle body (i.e. the spatially extended vehicle is considered), the admittance function has to be applied. As a result, Equation (3.35) becomes

$$\begin{aligned} S_{F'F'}(f) &= \chi_F^2(f)(\rho_L A_C C_F u_0 (1 + \frac{1}{2C_F}\frac{\mathrm{d}C_F}{\mathrm{d}\beta_\omega}\cot\beta_\omega))^2 S_{u'u'}(f) \\ &= (\rho_L A_C C_F u_0 (1 + \frac{1}{2C_F}\frac{\mathrm{d}C_F}{\mathrm{d}\beta_\omega}\cot\beta_\omega))^2 \tilde{S}_{u'u'}(f) \end{aligned} \tag{3.36}$$

where

$$\tilde{S}_{u'u'}(f) = \chi_F^2(f) S_{u'u'}(f) \tag{3.37}$$

Here, the equivalent wind spectrum $\tilde{S}_{u'u'}(f)$ which considers the actual effect of wind turbulence on the vehicle has been introduced.

Based on Taylor's frozen turbulence hypothesis, Cooper derived the wind spectrum relative to a moving vehicle as follows [59] [202]:

$$\frac{fS_{u'u'}(f)}{\sigma_{u'}^2} = \left[\frac{4\tilde{f}}{(1+70.8\tilde{f}^2)^{5/6}}\right]\left[\left(\frac{u_0+v_0\sin\alpha_\omega}{\bar{V}_{res}}\right)^2 + \left(1-\left(\frac{u_0+v_0\sin\alpha_\omega}{\bar{V}_{res}}\right)^2\right)\frac{0.5+94.4\tilde{f}^2}{1+70.8\tilde{f}^2}\right] \tag{3.38}$$

where f is the frequency, $\sigma_{u'}^2$ is the variance of wind turbulence and $\tilde{f}$ is the normalized frequency which is given by

$$\tilde{f} = \frac{fL_{u'}}{\bar{V}_{res}} \tag{3.39}$$

and

$$L_{u'} \approx L_{u'x}\sqrt{\left(\frac{u_0+v_0\sin\alpha_\omega}{\bar{V}_{res}}\right)^2 + 0.49\left(1-\left(\frac{u_0+v_0\sin\alpha_\omega}{\bar{V}_{res}}\right)^2\right)} \tag{3.40}$$

$$L_{u'x} = 50\frac{z^{0.35}}{z^{0.063}} \tag{3.41}$$

Combining Equation (3.37) and (3.38), the equivalent wind PSD for a moving vehicle can be expressed as follows

$$\tilde{S}_{u'u'}(f) = \chi_F^2(f)S_{u'u'}(f) = \chi_F^2(f)\left[\frac{4L_{u'}\sigma_{u'}^2}{\bar{V}_{res}(1+70.8\tilde{f}^2)^{5/6}}\right] \cdot \left[\left(\frac{u_0+v_0\sin\alpha_\omega}{\bar{V}_{res}}\right)^2 + \left(1-\left(\frac{u_0+v_0\sin\alpha_\omega}{\bar{V}_{res}}\right)^2\right)\frac{0.5+94.4\tilde{f}^2}{1+70.8\tilde{f}^2}\right] \tag{3.42}$$

Figure 3.4 shows the shapes of PSD for the turbulent wind acting on a moving vehicle with different vehicle velocities. As can be seen, the vehicle experiences higher levels of wind energy at higher vehicle speed. In addition, with the increase of vehicle speed, the shapes of the PSD changes only a little, but it moves quickly to the high frequency range, which implies that the turbulence may cause significant excitation on the vehicle at high vehicle speed [59].

When the admittance function is considered, the equivalent PSD at higher frequency becomes smaller due to the influence of the aerodynamic admittance. This can be seen in Figure 3.5.

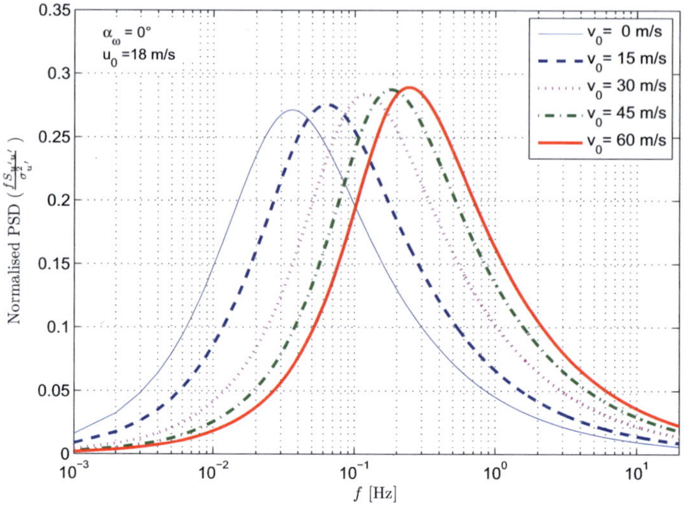

Figure 3.4: Power spectral density of wind relative to a moving vehicle with different vehicle speeds

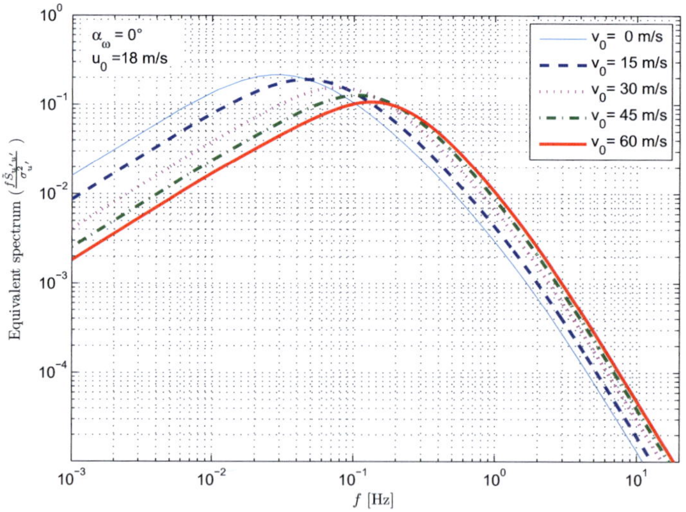

Figure 3.5: Equivalent spectrum of wind relative to a moving vehicle with different vehicle speeds

Based on Equations (3.36) and (3.42), the final spectrum for the fluctuating wind force of a moving vehicle can be obtained by

$$S_{F'F'}(f) = (\rho_L A_C C_F u_0 (1 + \frac{1}{2C_F}\frac{\mathrm{d}C_F}{\mathrm{d}\beta_\omega}\cot\beta_\omega))^2 \chi_F^2(f)$$
$$\cdot \left[\frac{4L_{u'}\sigma_{u'}^2}{\bar{V}_{res}(1+70.8\tilde{f}^2)^{5/6}}\right]\left[\left(\frac{u_0+v_0\sin\alpha_\omega}{\bar{V}_{res}}\right)^2 + \left(1-\left(\frac{u_0+v_0\sin\alpha_\omega}{\bar{V}_{res}}\right)^2\right)\frac{0.5+94.4\tilde{f}^2}{1+70.8\tilde{f}^2}\right] \tag{3.43}$$

An equivalent expression of Equation (3.43) in the time domain can be expressed as follows

$$F' = \left(\rho_L A_C C_F u_0 (1 + \frac{1}{2C_F}\frac{\mathrm{d}C_F}{\mathrm{d}\beta_\omega}\cot\beta_\omega)\right)^2 \int_0^t h_F(t-\tau)\, u'(\tau)\mathrm{d}\tau \tag{3.44}$$

where $h_F(t-\tau)$ refers to the weighting function.

Spatially extended vehicles with varying vehicle speed

Now, considering a general case, i.e. the vehicle running with a varying vehicle speed $v(t)$ under the excitation of wind with turbulence, the total wind excitations can be ideally written as follows

$$\begin{aligned} F(t) &= \frac{1}{2}\rho_L A_C C_F(\beta_\omega(t)) V_{res}^2(t) \\ &= \frac{1}{2}\rho_L A_C C_F(\beta_\omega(t))\left(v(t)^2 + u(t)^2 + 2u(t)v(t)\sin(\alpha_w)\right) \end{aligned} \tag{3.45}$$

$$\begin{aligned} M(t) &= \frac{1}{2}\rho_L A_C L_C C_M(\beta_\omega(t)) V_{res}^2(t) \\ &= \frac{1}{2}\rho_L A_C L_C C_M(\beta_\omega(t))\left(v(t)^2 + u(t)^2 + 2u(t)v(t)\sin(\alpha_w)\right) \end{aligned} \tag{3.46}$$

where $F(t)$ and $M(t)$ refers to the nonatationary wind force and moments, respectively. $C_M(\beta_\omega(t))$ is the aerodynamic coefficient of the wind moment.

Note that the above equations are only valid for a stationary case. They have not taken the spatial correlation of the wind turbulence into account. Consequently, the admittance function has to be adopted to calculate the unsteady components of the aerodynamic excitations and the corresponding relative wind speed has to be changed as follows [15] [39]:

$$\begin{aligned} V_{res}^2(t) = V_{res\_c}^2(t) &= (v(t) + (u_0 + u'_c(t))\sin(\alpha_\omega))^2 + ((u_0 + u'_c(t))\cos(\alpha_\omega))^2 \\ &= v^2(t) + (u_0 + u'_c(t))^2 + 2v(t)\,(u_0 + u'_c(t))\sin(\alpha_\omega)\cos(\alpha_\omega) \end{aligned} \tag{3.47}$$

where $V_{res\_c}(t)$ refers to the corrected relative wind speed and $u'_c(t)$ refers to the wind turbulence obtained based on the following equivalent spectrum

$$\tilde{S}_{u'u'}(f) = \chi_F^2(f)\left[\frac{4L_{u'}\sigma_{u'}^2}{\bar{V}_{res}(1+70.8\tilde{f}^2)^{5/6}}\right]\cdot\left[C_u^2 + \left(1 - C_u^2\right)\frac{0.5+94.4\tilde{f}^2}{1+70.8\tilde{f}^2}\right] \tag{3.48}$$

where

$$C_u^2 = \frac{\left(u_0\Delta t + \int_0^t v(\tau)\sin(\alpha_\omega)d\tau\right)^2}{\left(u_0\Delta t + \int_0^t v(\tau)\sin(\alpha_\omega)d\tau\right)^2 + \left(\int_0^t v(\tau)\cos(\alpha_\omega)d\tau\right)^2} \tag{3.49}$$

In particular, for a vehicle accelerating or decelerating with a constant acceleration a_c,

$$C_u^2 = \frac{\left(u_0\Delta t + (v_0t + \frac{1}{2}a_ct^2)\sin(\alpha_\omega)\right)^2}{\left(u_0\Delta t + (v_0t + \frac{1}{2}a_ct^2)\sin(\alpha_\omega)\right)^2 + \left((v_0t + \frac{1}{2}a_ct^2)\cos(\alpha_\omega)\right)^2} \tag{3.50}$$

3.1.4 Vortex shedding and buffeting

Wind excitations on the vehicle are usually caused by vortex shedding and buffeting [11]. For a large vehicle running through a certain scenario (e.g. bridge abutment, a gap in a hedge), the effect of the wind excitation may be exaggerated by the buffeting from the unsteady wake around the vehicle [171].

In the last few decades, a great number of computational simulations as well as wind tunnel tests have been implemented to study the flow structure around the high-speed vehicles in crosswinds. It has been found that the dominant flow pattern was a series of inclined vortex. A more conventional vortex shedding pattern can be seen at higher yaw angles. Besides, at intermediate angles there are certain switchings of the flow [13]. Figure 3.6 depicts a sketch for the flow pattern of the crosswind around the train. As can be seen from the figure, vortex tubes occur frequently in the leeward side of the vehicle. Chiu and Squire [51] simplified the train as a slender body and carried out the experiments in a closed wind tunnel. It was found that there were three regimes as the yaw angle increases. The flow is similar as the steady flow at small yaw angles up to about 45° and then with the increase of the yaw angle, the flow becomes unsteady. When the yaw angle is larger than 60°, the flow will become totally unsteady. Conclusions also show that the distance from the nose where vortex shedding begins to occur is smaller at larger yaw angles. The larger the yaw angle is, the stronger the corresponding vortex shedding appears. A schematic diagram for the relationship between the flow nature and the wind yaw angle can be seen in Figure 3.7. Based on the similar slender body with wheels, Copley [60] developed a numerical method which can predict the aerodynamic force on a train with reasonable accuracy for small yaw angles. The flows far from the nose are suggested to be treated as a two-dimensional scheme due to the marginally different stream line pattern. Chiu [49] developed a two-dimensional second-order vortex panel method for the calculation of wind

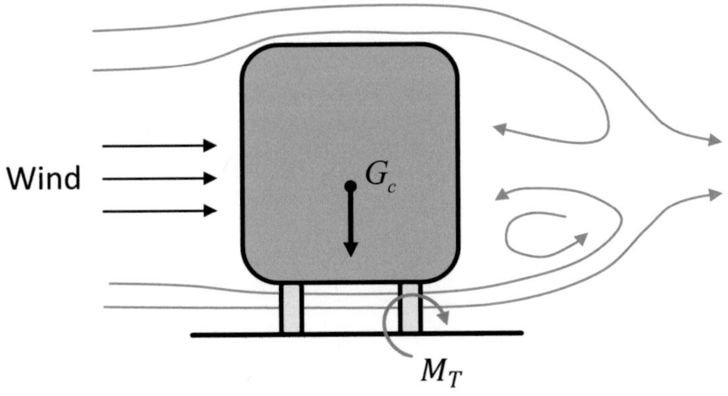

Figure 3.6: Sketch for the flow pattern of the crosswind around train

forces and moments away from the train nose. Recently, more and more CFD methods like Reynolds-averaged Naiver-Stokes (RANS) and large-eddy simulation (LES) are developed for the prediction of the crosswind forces acting on the vehicles [46] [96] [73]. There is a broad agreement between the wind tunnel tests and the CFD results that the higher the yaw angle is, the more conventional vortex shedding pattern will be observed.

Studies have shown that the wind force induced by the vortex shedding has a narrow-band frequency which concentrates around the Strouhal frequency f_{st} [53] [51] [13]

$$f_{st} = \frac{u(t)S_t}{L} \tag{3.51}$$

where $u(t)$ means the wind velocity, L refers to the characteristic length of the cross-section at which the resonant vortex shedding occurs. S_t is the Strouhal number that describes the oscillating flow mechanisms and can be obtained by the experimental method [121] [172]. Hemida [96] indicates that there are two dominating modes of the wake unsteadiness around the train due to the excitation of the crosswind, that is, the wake oscillation with a Strouhal number of 0.1 in the horizontal plane and a relatively weaker vortex shedding motion with the Strouhal number of 0.15 to 0.20.

Assume that the train's cross-section shape is approximately rectangular, then the Strouhal number for the train in a strong crosswind can be easily obtained, see Figure 3.8. As a result of the vortex shedding, the alternating force which is perpendicular to the flow will arise.

For occasions like the train running on the bridge with strong turbulence, buffeting phenomenon may happen. A classical quasi-steady model for the wind side/lifting forces induced by the buffeting phenomenon can be given as follows [52]:

$$\begin{bmatrix} F_S(t) \\ F_L(t) \end{bmatrix} = \frac{1}{2}\rho_L \cdot V_{res}^2(t) \cdot A_C \cdot \begin{bmatrix} \cos(\beta'_\omega) & -\sin(\beta'_\omega) \\ \sin(\beta'_\omega) & \cos(\beta'_\omega) \end{bmatrix} \begin{bmatrix} C_S(\beta_\omega(t)) \\ C_L(\beta_\omega(t)) \end{bmatrix} \tag{3.52}$$

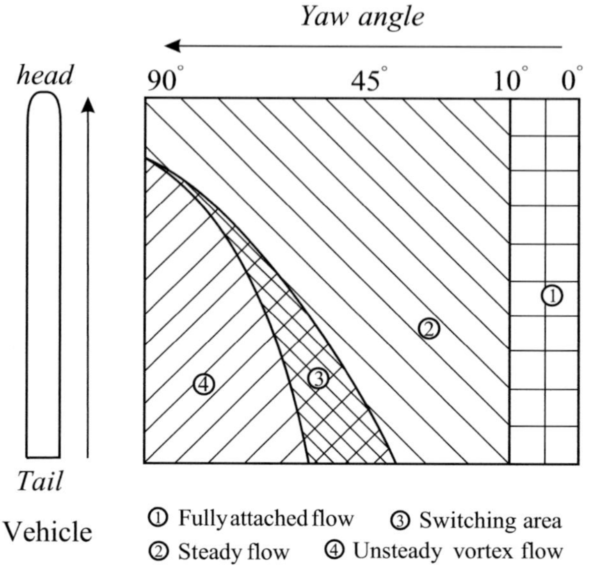

Figure 3.7: Variation of the flow pattern vs. wind yaw angle [51]

$$V_{res}(t) = \sqrt{\left(u_0 + u'(t) - \dot{\tilde{y}}_l(t)\right)^2 + \left(u'_w(t) - \dot{\tilde{y}}_v(t)\right)^2} \tag{3.53}$$

where $u'(t)$ and $u'_w(t)$ are the turbulent wind velocity in the lateral and vertical directions. $\dot{\tilde{y}}_l$ and $\dot{\tilde{y}}_v$ refer to the corresponding train velocity in the wind coordinate system. By adding the train velocity, the aeroelastic effects related to the vibrations of the vehicle have been taken into account. However, in practical use, $\dot{\tilde{y}}_l$ and $\dot{\tilde{y}}_v$ are usually ignored since they are usually very small.

Assume that the wind turbulence comes mainly from the mean wind speed direction, i.e. the vertical wind turbulence $u'_w(t) = 0$ and the change of the relative wind angle caused by the turbulence is negligible, then Equation (3.52) can be simplified to Equation (3.54), which shows the same characteristic of the wind force as before, cf. Equation (3.9).

$$\begin{bmatrix} F_S(t) \\ F_L(t) \end{bmatrix} = \frac{1}{2}\rho_L \cdot V_{res}^2(t) \cdot A_C \cdot \begin{bmatrix} 1 & 0 \\ 0 & 1 \end{bmatrix} \begin{bmatrix} C_S(\beta_\omega(t)) \\ C_L(\beta_\omega(t)) \end{bmatrix} \tag{3.54}$$

For unsteady wind force in the time domain, the aerodynamic weighting function $h_F(\tau)$ have to be applied, thus Equation (3.52) can be changed to

$$\begin{aligned} F_S = & \frac{1}{2} \cdot \rho_L \cdot A_C \cdot \int_0^t h_{s,u}(t-\tau) \cdot V_{res}^2(\tau) \cdot C_S(\beta_\omega(\tau)) \cos({\beta'}_\omega(\tau)) \mathrm{d}\tau \\ & - \frac{1}{2} \cdot \rho_L \cdot A_C \cdot \int_0^t h_{s,w}(t-\tau) \cdot V_{res}^2(\tau) \cdot C_L(\beta_\omega(\tau)) \sin({\beta'}_\omega(\tau)) \mathrm{d}\tau \end{aligned} \tag{3.55}$$

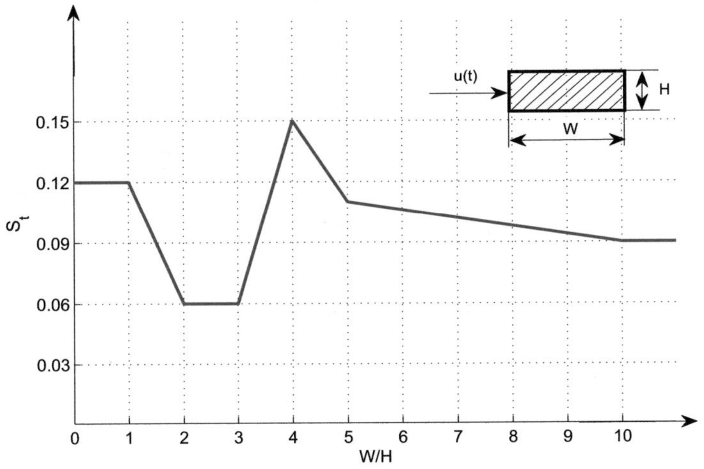

Figure 3.8: Strouhal number for rectangular cross-sections [81]

$$
\begin{aligned}
F_L = & \frac{1}{2} \cdot \rho_L \cdot A_C \cdot \int_0^t h_{l,u}(t-\tau) \cdot V_{res}^2(\tau) \cdot C_L(\beta_\omega(\tau)) \cos({\beta'}_\omega(\tau)) \mathrm{d}\tau \\
& - \frac{1}{2} \cdot \rho_L \cdot A_C \cdot \int_0^t h_{l,w}(t-\tau) \cdot V_{res}^2(\tau) \cdot C_S(\beta_\omega(\tau)) \sin({\beta'}_\omega(\tau)) \mathrm{d}\tau
\end{aligned}
\tag{3.56}
$$

where $h_{s,u}, h_{s,w}, h_{l,u}, h_{l,w}$ are the aerodynamic weighting functions of the side force and lift force in the lateral and vertical directions, respectively. When the vertical wind turbulence $u'_w(t)$ is omitted, $h_{s,w}$ and $h_{l,w}$ will be set as zero.

3.2 HHT analysis of the wind excitation

As a traditional signal-processing tool, Fourier transform (FT) has been widely used to analyze the random signals in the past. FT is quite useful for signals whose frequency content is stationary. However, for frequency modulated signals or other nonstationary signals, FT often fails to reveal the time-varying character of the signal [109]. Later, the wavelet technique was developed, although it is capable for analyzing the nonlinear and nonstationary signals, many deficiencies (e.g. overlapping in adjacent scale and distortion of the signal) have been reported when using it, cf. [198]. To circumvent these problems, the emerged Hilbert-Huang transform (HHT) method [103] is applied here.

The HHT method is composed of empirical mode decomposition (EMD) and Hilbert transform (HT) [154]. This means when the HHT method is used, the first step is to decompose the original signal into a series of intrinsic mode functions (IMF) by EMD. Then the Hilbert transform will be applied. The essence of EMD is to identify the intrinsic oscillatory modes by their characteristic time scales in the data [103]. It decomposes the signal data into corresponding IMF via the sifting process.

Studies have shown that EMD is quite versatile and has a wide application for extracting signals from data that generated by nonlinear or non-stationary processes (e.g. [102]). It can reduce any data into simple independent IMF, whether they are linear, nonlinear or nonstationary [217]. However, the original EMD has a drawback, that is, the frequent appearance of mode mixing. To avoid this problem, a method named ensemble empirical mode decomposition (EEMD) with the assistance of adding white noises is then proposed [209].

The basic idea of EEMD is to define the actual IMF components as a mean of the ensemble of trials, each composing the signal plus a white noise with finite amplitude. Although the individual trial may produce many noisy results due to the noise-added decompositions, the noise will be canceled out in the ensemble mean of enough trails. In other words, finally, as more and more trails are added in the ensemble, the only persistent part is the signal without noise.

Suppose that $x(t)$ is the original nonstationary signal, $c_i(t)$ is the component of each IMF, $r(t)$ is the final residue which reflects the trend of the signal. Then the original time signal can be decomposed by the following equation [104]:

$$x(t) = \sum_{i=1}^{n} c_i(t) + r(t) \tag{3.57}$$

where n is the total number of IMF.

After obtaining each IMF component from the time history $x(t)$, the HT will be applied so that the instantaneous frequency of each IMF can be obtained. For an arbitrary $x(t)$, its Hilbert transform $\tilde{x}(t)$ is defined as [105]:

$$\tilde{x}(t) = \mathrm{H}\left[x(t)\right] = \frac{1}{\pi} P \int_{-\infty}^{\infty} \frac{x(\tau)}{t-\tau}\,\mathrm{d}\tau \tag{3.58}$$

where H denotes the HT operator, P is the Cauchy principal value.

Coupling $x(t)$ and $\tilde{x}(t)$, an analytical signal $z(t)$ of $x(t)$ can be obtained:

$$z(t) = x(t) + i\tilde{x}(t) = A(t)e^{i\theta(t)} \tag{3.59}$$

$$A(t) = \left[x^2(t) + \tilde{x}^2(t)\right]^{1/2}, \ \theta(t) = \arctan\left(\frac{\tilde{x}(t)}{x(t)}\right) \tag{3.60}$$

where $A(t)$ is the instantaneous amplitude of $x(t)$, which reflects the energy variation of x(t) in the time domain. $\theta(t)$ is the corresponding instantaneous phase angle. The instantaneous frequency $\omega(t)$ is defined as a derivative of the instantaneous phase angle $\theta(t)$

$$\omega(t) = \frac{\mathrm{d}\theta(t)}{\mathrm{d}t} \tag{3.61}$$

Now, the original signal $x(t)$ without the final residue can be defined by the instantaneous amplitude $A(t)$ and phase angle $\theta(t)$ [101]

$$x(t) = \sum_{i=1}^{n} A_i(t) \cos\left(\theta_i(t)\right) = \sum_{i=1}^{n} A_i(t) \cos\left(\int_0^t \omega_i(t)\mathrm{d}t\right) \tag{3.62}$$

In order to analyze the signal in a time-frequency domain, Huang et al. [101] [103] introduced the Hilbert spectrum which can represent the distribution of the energy (amplitude) and frequency with time

$$H(\omega, t) = \sum_{j=1}^{n} A_j(t)\ \exp(i \int_0^t \omega_j(t)\ \mathrm{d}t) \tag{3.63}$$

In addition, Huang et al.[103] also introduced the marginal spectrum $h(\omega)$ as

$$h(\omega) = \int_0^T H(\omega, t)\,\mathrm{d}t \tag{3.64}$$

It should be pointed out that the frequency in either $H(\omega, t)$ or $h(\omega, t)$ has a total different meaning from the traditional Fourier spectral analysis [103]. In the Fourier spectrum, a certain frequency ω means that there is a sine or cosine wave existing in the whole time span of the data. However, in Hilbert spectrum, it only indicates that there is a higher likelihood for such a wave to appear. Accordingly, the frequency in $h(\omega)$ only shows the likelihood that an oscillation with such a frequency exists.

To analyze nonstationary characteristics of the wind excitation, wind excitations caused by the nonstationary gust are studied. Figure 3.9 represents the original wind force produced by the nonstationary gust and all of the components (IMFs and residue) derived by EEMD. There are total nine components, i.e. C1 $\sim$ C9, among which C1 $\sim$ C8 refer to the IMFs while C9 refers to the residue. By summing up different IMFs, the trend of the signal in the time scale and the process of the EEMD can be clearly seen, cf. Figure 3.10.

The instantaneous frequency of the IMF decomposed from the nonstationary crosswind force can be seen in Figure 3.11 (here, taking IMF_4 (C4) for example). It can be seen that the frequency of IMF_4 changes time by time, which indicates strong nonstationary characteristics of the gust.

Figure 3.12 and 3.13 show an example of the time-frequency diagram for the nonstationary gust and the corresponding wind force, respectively. As can be seen from the figure, the frequencies of the gust and the wind force distribute over all the time span. They change in every instant of time. At about $t = 17$ s, the wind energy becomes stronger since the gust achieves its amplitude.

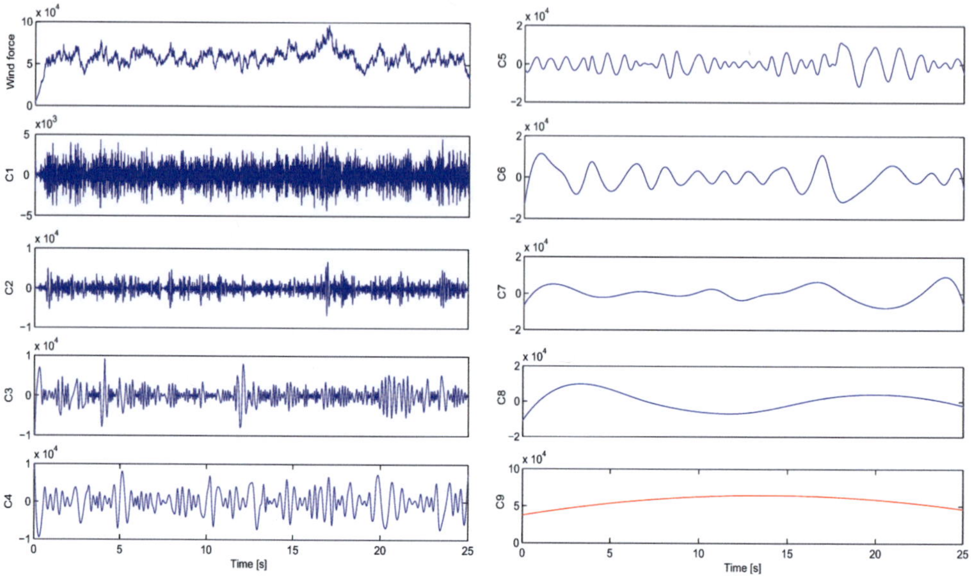

(a) Wind force signal and the decomposition components C1 ~ C4

(b) The decomposition components C5 ~ C9

Figure 3.9: The original wind force and the corresponding decomposition components

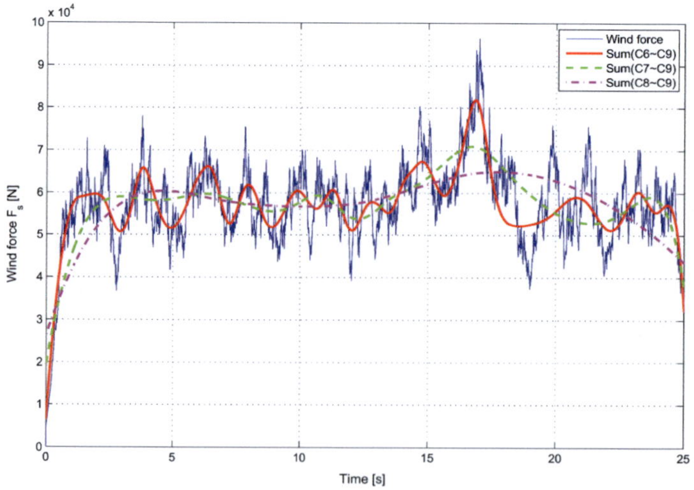

Figure 3.10: Trends of the wind force produced by nonstationary gust

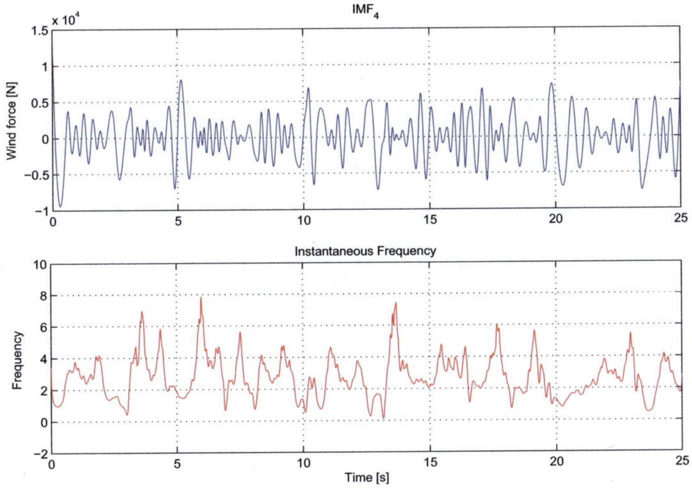

Figure 3.11: Instantaneous frequency of IMF_4.

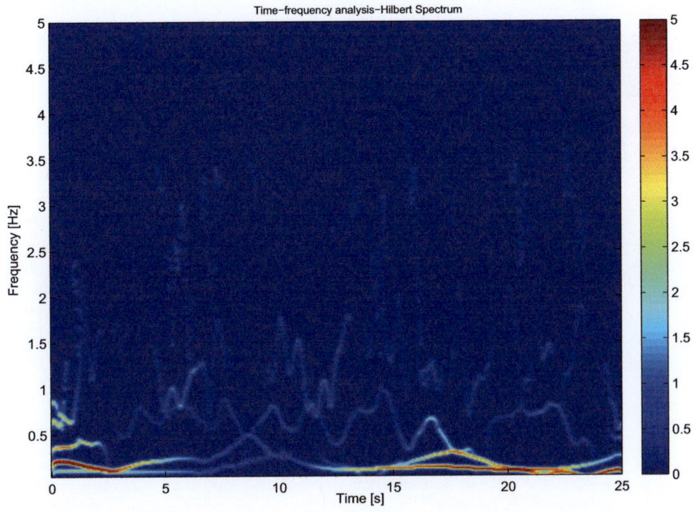

Figure 3.12: Time-frequency analysis for the gust with nonstationary turbulence (u_0 = 18 m/s)

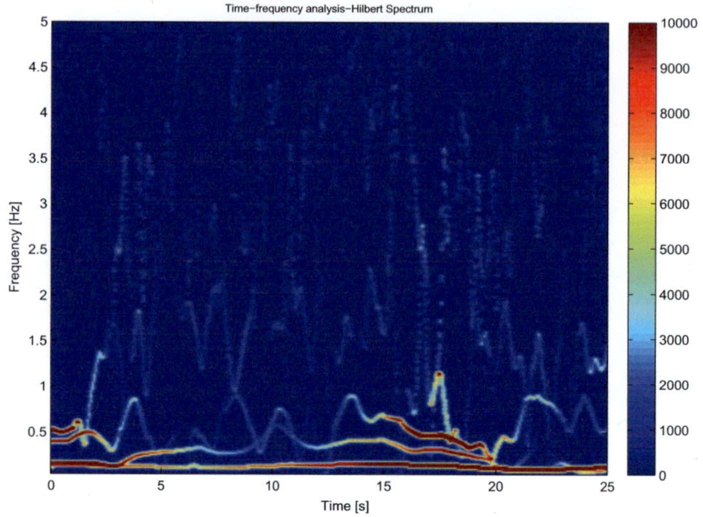

Figure 3.13: Time-frequency analysis for the nonstationary wind force on a moving vehicle with constant vehicle speed ($u_0 = 18$ m/s, $v_0 = 160$ km/h)

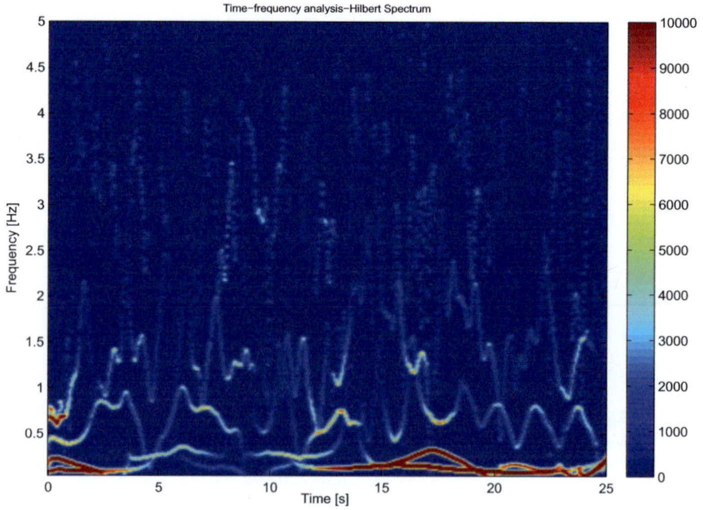

Figure 3.14: Time-frequency analysis for the nonstationary wind force on a moving vehicle with acceleration a_c ($u_0 = 18$ m/s, $v_0 = 160$ km/h, $a_c = 1$ m/s^2)

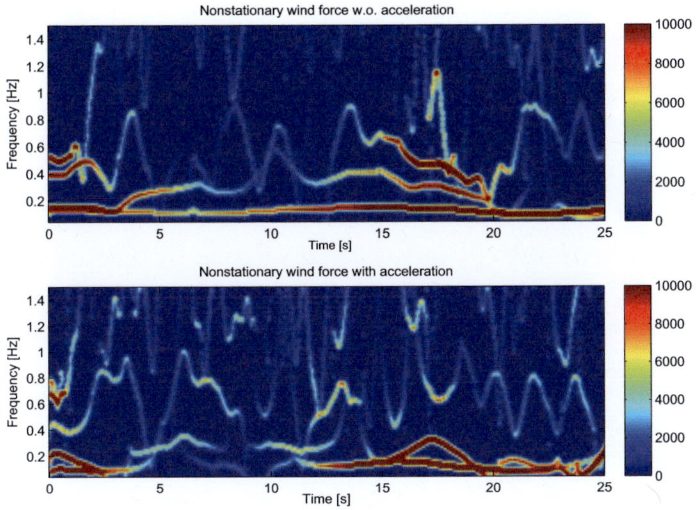

Figure 3.15: Comparison of the Hilbert spectrum for the nonstationary wind force on a moving vehicle w. or w.o. acceleration

For comparison, the EEMD and HHT analysis of the nonstationary wind force acting on a moving vehicle with acceleration has also been represented, see Figure 3.14. Compared to Figure 3.13, the frequency of Hilbert spectrum in Figure 3.14 changes faster and more intensively. In Figure 3.13, the main energy of the spectrum is concentrated in $0 \sim 1$ Hz while in Figure 3.14 the main energy is concentrated in $0 \sim 1.5$ Hz.

A much clearer comparison for the HHT spectrum of nonstationary wind force on the vehicle with or without acceleration can be seen in Figure 3.15, which shows the spectrum in a relatively smaller frequency range. It shows again that the acceleration of the vehicle makes the wind energy move to a higher frequency range, which is in a good agreement with the earlier study by Cooper [59], cf. Figure 3.4.

4 Modeling and simulation of the vehicle system

This chapter mainly introduces the modeling process of the vehicle system. Both road and railway vehicles are considered. The road vehicles are classified into several types and modeled in MATLAB/Simulink while the railway vehicles are simulated with the commercial multibody system code (MBS-Code) ADAMS/Rail.

The aim of this chapter is to model the vehicles running in strong crosswind with different vehicle speeds, including other conditions (e.g. road/track conditions, driver models for road vehicles, different wind scenarios, etc.), and hence to establish a basis for the simulation of crosswind stability.

The integrated vehicle-wind system is composed of several subsystems, that is, the nonstationary crosswind, the nonlinear vehicle and the topographic and environmental scenarios (e.g. road vehicles running on wet road or railway vehicles running on curved track). Nonstationary crosswind has already been studied and discussed before. In the following part, the related contents about the modeling of nonlinear vehicle system as well as the different scenarios will be presented.

4.1 Road vehicle

It has been noted that the influence of crosswind on different types of road vehicles varies widely. Therefore, to study the crosswind stability of road vehicles, vehicle types have to be identified.

As far as the vehicle types are concerned, following types of vehicles are considered in this work:

1) Passenger car

2) Truck

3) Bus

4) Van

5) Articulated car-trailer

Except for the car-trailer, 2-axle, 4-wheel vehicle models are adopted. Even if the vehicles have more than two axles, they can be transformed into equivalent 2-axle vehicles for the

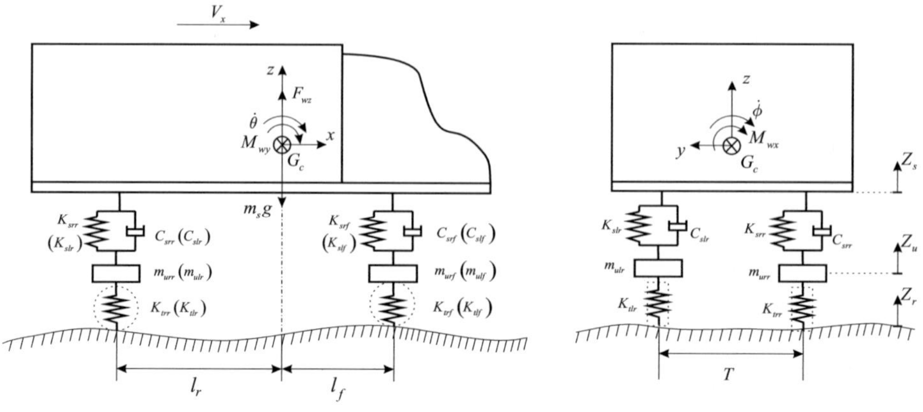

Figure 4.1: Truck model for crosswind stability analysis

dynamic analysis [45]. For the car-trailer, the vehicle model is considered to be composed of two parts, namely the car and the trailer, which are combined by a hitch force. The detailed description for the car-trailer can be seen in Section 4.1.5.

The vehicles are modeled as a combination of rigid bodies with lumped masses as well as springs and dampers which represent the suspension characteristics of the vehicle. Both the horizontal and vertical movements of the vehicle are considered. Masses of the wheels and suspensions are assumed to be concentrated. The springs and dampers are massless. To obtain a conservative assessment of the crosswind stability, unladen vehicles, which are most prone to wind-induced accidents, have been considered.

In addition, due to the large influence of the tire forces on vehicle dynamics, a tire model has to be considered. Strong crosswinds affect the state of the vehicle and thus the tire forces.

Finally, road conditions (dry/wet) and driver models are all taken into account for the simulation. The influence of road conditions and driver models on the vehicle dynamics are studied. According to [48], the friction coefficients of the road are set as 0.9 for dry conditions and 0.5 the wet conditions.

4.1.1 Vehicle model with two axles and four wheels

In this section, mathematical models for vehicle dynamics as well as the coupled vehicle-wind system are described. Nonlinear vehicle models are studied and developed based on foregoing studies, see [32] [178] [110] [180] and [107].

Take a truck for example, cf. Figure 4.1. The wind excitations acting on the vehicle are considered as concentrated forces and moments in three directions, i.e. the longitudinal, lateral and vertical directions, respectively. When the wind forces are taken into account, the force-balance equation of the sprung mass can be described as follows:

$$F_{slf} + F_{srf} + F_{slr} + F_{srr} + F_{wz} = m_s \ddot{Z}_s \tag{4.1}$$

where F_{slf}, F_{srf}, F_{slr}, F_{srr} are the suspension force at left-front, right-front, left-rear and right-rear corner in vertical direction, respectively. F_{wz} refers to the wind force in z direction. m_s is the sprung mass of the vehicle and Z_s is the displacement of the sprung mass at the center of gravity.

The suspension forces at each corner can be defined by spring forces and damping forces as follows:

$$\begin{bmatrix} F_{slf} \\ F_{srf} \\ F_{slr} \\ F_{srr} \end{bmatrix} = \begin{bmatrix} K_{slf} & 0 & 0 & 0 \\ 0 & K_{srf} & 0 & 0 \\ 0 & 0 & K_{slr} & 0 \\ 0 & 0 & 0 & K_{srr} \end{bmatrix} \begin{bmatrix} Z_{ulf} - Z_{slf} \\ Z_{urf} - Z_{srf} \\ Z_{ulr} - Z_{slr} \\ Z_{urr} - Z_{srr} \end{bmatrix} + \begin{bmatrix} C_{slf} & 0 & 0 & 0 \\ 0 & C_{srf} & 0 & 0 \\ 0 & 0 & C_{slr} & 0 \\ 0 & 0 & 0 & C_{srr} \end{bmatrix} \begin{bmatrix} \dot{Z}_{ulf} - \dot{Z}_{slf} \\ \dot{Z}_{urf} - \dot{Z}_{srf} \\ \dot{Z}_{ulr} - \dot{Z}_{slr} \\ \dot{Z}_{urr} - \dot{Z}_{srr} \end{bmatrix} \tag{4.2}$$

where K_{si} and C_{si} are the corresponding spring stiffness and damping coefficient of the suspension, $i = lf, rf, lr, rr$ refers to the left-front, right-front, left-rear and right-rear position, respectively. Same subscript for i will be also used in the following part of this section. Consequently, Z_{ui}, Z_{si}, $\dot{Z}_{ui}$, $\dot{Z}_{si}$ are the displacement and velocity of the unsprung and sprung mass at each corner.

The position of the sprung mass at each corner can be determined by the sprung mass displacement Z_s, the pitch angle θ and the roll angle ϕ:

$$\begin{bmatrix} Z_{slf} \\ Z_{srf} \\ Z_{slr} \\ Z_{srr} \end{bmatrix} = \begin{bmatrix} 1 & -l_f & T/2 \\ 1 & -l_f & -T/2 \\ 1 & l_r & T/2 \\ 1 & l_r & -T/2 \end{bmatrix} \begin{bmatrix} Z_s \\ \sin\theta \\ \sin\phi \end{bmatrix} \tag{4.3}$$

where l_f is the distance from the front axle to the gravity center, l_r is the distance from the rear axle to the gravity center and T is the track width.

Assume that the pitch angle θ and the roll angle ϕ are small, Equation (4.3) can be simplified to:

$$\begin{bmatrix} Z_{slf} \\ Z_{srf} \\ Z_{slr} \\ Z_{srr} \end{bmatrix} = \begin{bmatrix} 1 & -l_f & T/2 \\ 1 & -l_f & -T/2 \\ 1 & l_r & T/2 \\ 1 & l_r & -T/2 \end{bmatrix} \begin{bmatrix} Z_s \\ \theta \\ \phi \end{bmatrix} \tag{4.4}$$

The unsprung position at each corner can be obtained from:

$$\begin{bmatrix} m_{ulf} & 0 & 0 & 0 \\ 0 & m_{urf} & 0 & 0 \\ 0 & 0 & m_{ulr} & 0 \\ 0 & 0 & 0 & m_{urr} \end{bmatrix} \begin{bmatrix} \ddot{Z}_{ulf} \\ \ddot{Z}_{urf} \\ \ddot{Z}_{ulr} \\ \ddot{Z}_{urr} \end{bmatrix} = K_t \begin{bmatrix} Z_{rlf} \\ Z_{rrf} \\ Z_{rlr} \\ Z_{rrr} \end{bmatrix}$$
$$+ \begin{bmatrix} K_{slf} & C_{slf} \\ K_{srf} & C_{srf} \\ K_{slr} & C_{slr} \\ K_{srr} & C_{srr} \end{bmatrix} \begin{bmatrix} Z_s \\ \dot{Z}_s \end{bmatrix} + \begin{bmatrix} -l_f K_{slf} & -l_f C_{slf} & K_{slf}T/2 & C_{slf}T/2 \\ -l_f K_{srf} & -l_f C_{srf} & -K_{srf}T/2 & -C_{srf}T/2 \\ l_r K_{slr} & l_r C_{slr} & K_{slr}T/2 & C_{slr}T/2 \\ l_r K_{srr} & l_r C_{srr} & -K_{srr}T/2 & -C_{srr}T/2 \end{bmatrix} \begin{bmatrix} \theta \\ \dot{\theta} \\ \phi \\ \dot{\phi} \end{bmatrix}$$
$$+ \begin{bmatrix} 1 & 0 & 0 & 0 \\ 0 & 1 & 0 & 0 \\ 0 & 0 & 1 & 0 \\ 0 & 0 & 0 & 1 \end{bmatrix} \begin{bmatrix} -(K_{slf}+K_t) \\ -(K_{srf}+K_t) \\ -(K_{slr}+K_t) \\ -(K_{srr}+K_t) \end{bmatrix} \begin{bmatrix} Z_{ulf} \\ Z_{urf} \\ Z_{ulr} \\ Z_{urr} \end{bmatrix} + \begin{bmatrix} -C_{slf} & 0 & 0 & 0 \\ 0 & -C_{srf} & 0 & 0 \\ 0 & 0 & -C_{slr} & 0 \\ 0 & 0 & 0 & -C_{srr} \end{bmatrix} \begin{bmatrix} \dot{Z}_{ulf} \\ \dot{Z}_{urf} \\ \dot{Z}_{ulr} \\ \dot{Z}_{urr} \end{bmatrix} \tag{4.5}$$

Based on Equations (4.1), (4.2), (4.4) and (4.5), the acceleration of the sprung mass in z direction can be obtained as follows (under assumption that $K_{slf} = K_{srf}, K_{slr} = K_{srr}, C_{slf} = C_{srf}, C_{slf} = C_{srf}$):

$$\begin{aligned} m_s\ddot{Z}_s = {} & F_{wz} - (K_{slf}+K_{srf}+K_{slr}+K_{srr})\,Z_s - (C_{slf}+C_{srf}+C_{slr}+C_{srr})\,\dot{Z}_s \\ & + (K_{slf}l_f + K_{srf}l_f - K_{slr}l_r - K_{srr}l_r)\,\theta + (C_{slf}l_f + C_{srf}l_f - C_{slr}l_r - C_{srr}l_r)\,\dot{\theta} \\ & + K_{slf}Z_{ulf} + C_{slf}\dot{Z}_{ulf} + K_{srf}Z_{urf} + C_{srf}\dot{Z}_{urf} + K_{slr}Z_{ulr} + C_{slr}\dot{Z}_{ulr} + K_{srr}Z_{urr} + C_{srr}\dot{Z}_{urr} \end{aligned} \tag{4.6}$$

The moment balance equation of the roll movement about x axis for the vehicle are expressed by

$$\begin{aligned} I_{xx}\ddot{\phi} &= M_{wx} + \tfrac{1}{2}F_{slf}T - \tfrac{1}{2}F_{srf}T + \tfrac{1}{2}F_{slr}T + \tfrac{1}{2}F_{srr}T \\ &= M_{wx} - (K_{slf}+K_{srf}+K_{slr}+K_{srr})\,\tfrac{\phi T^2}{4} - (C_{slf}+C_{srf}+C_{slr}+C_{srr})\,\tfrac{\dot{\phi} T^2}{4} + \tfrac{K_{slf}Z_{ulf}T}{2} \\ &\quad + \tfrac{C_{slf}\dot{Z}_{ulf}T}{2} - \tfrac{K_{srf}Z_{urf}T}{2} - \tfrac{C_{srf}\dot{Z}_{urf}T}{2} + \tfrac{K_{slr}Z_{ulr}T}{2} + \tfrac{C_{slr}\dot{Z}_{ulr}T}{2} - \tfrac{K_{srr}Z_{urr}T}{2} - \tfrac{C_{srr}\dot{Z}_{urr}T}{2} \end{aligned} \tag{4.7}$$

Likewise, the moment balance equation for pitch movement about y axis is given as

$$\begin{aligned} I_{yy}\ddot{\theta} = {} & M_{wy} - F_{slf}l_f - F_{srf}l_f + F_{slr}l_r + F_{srr}l_r = (K_{slf}l_f + K_{srf}l_f - K_{slr}l_r - K_{srr}l_r)\,Z_s \\ & + (C_{slf}l_f + C_{srf}l_f - C_{slr}l_r - C_{srr}l_r)\,\dot{Z}_s - \left(K_{slf}l_f^2 + K_{srf}l_f^2 + K_{slr}l_r^2 + K_{srr}l_r^2\right)\theta \\ & - \left(C_{slf}l_f^2 + C_{srf}l_f^2 + C_{slr}l_r^2 + C_{srr}l_r^2\right)\dot{\theta} - K_{slf}Z_{ulf}l_f - C_{slf}\dot{Z}_{ulf}l_f - K_{srf}Z_{urf}l_f \\ & - C_{srf}\dot{Z}_{urf}l_f + K_{slr}Z_{ulr}l_r + C_{slr}\dot{Z}_{ulr}l_r + K_{srr}Z_{urr}l_r + C_{srr}\dot{Z}_{urr}l_r \end{aligned} \tag{4.8}$$

Figure 4.2 illustrates the vehicle handling model under wind excitation. It takes the movement of the vehicle body along the longitudinal axis x and lateral axis y as well as

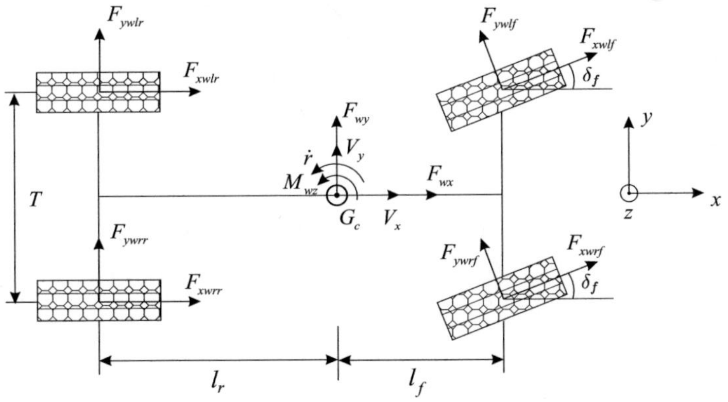

Figure 4.2: Vehicle handling model for the crosswind stability analysis

the yaw speed $\dot{r}$ into account. The steering angles of both front wheels are considered to be identical.

The longitudinal acceleration a_x and lateral acceleration a_y of the vehicle body can be determined by the following equations

$$a_x = \dot{V}_x - \dot{r}V_y \tag{4.9}$$

$$a_y = \dot{V}_y + \dot{r}V_x \tag{4.10}$$

where V_x, V_y are the vehicle velocity along x direction and y direction in the global coordinate system, respectively. By applying Newton's second law of motion, the balance equations of longitudinal and lateral forces of the vehicle can be described as:

$$ma_x = F_{xlf} + F_{xrf} + F_{xlr} + F_{xrr} + F_{wx} \tag{4.11}$$

$$ma_y = F_{ylf} + F_{yrf} + F_{ylr} + F_{yrr} + F_{wy} \tag{4.12}$$

where F_{xi}, F_{yi}, $i = lf,\ rf,\ lr,\ rr$ are the respective tire forces in x and y direction.

Figure 4.3 presents the relationship between the wheel-fixed coordinate system $x_w y_w z_w$ and the vehicle coordinate system xyz. The wheel force in the vehicle coordinate system is given as follows:

$$F_{xi} = F_{xwi}\cos\delta_i - F_{ywi}\sin\delta_i \tag{4.13}$$

$$F_{yi} = F_{xwi}\sin\delta_i + F_{ywi}\cos\delta_i \tag{4.14}$$

in which δ_i is the wheel steering angle, F_{xwi}, F_{ywi} are the longitudinal and lateral tire force in the wheel plane.

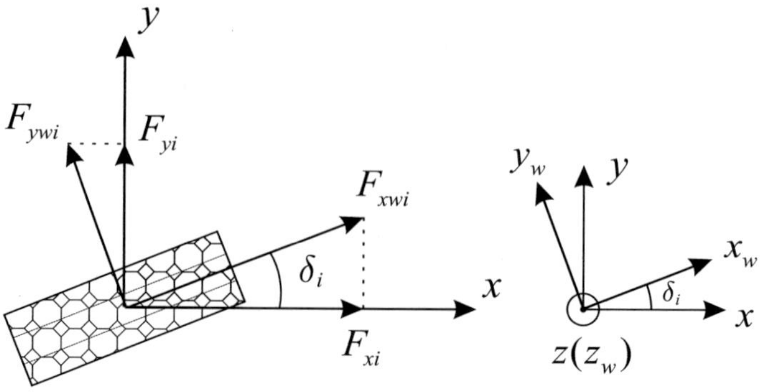

Figure 4.3: Wheel force in the vehicle coordinate system xyz

Assume that

$$\delta_{lf} = \delta_{rf} = \delta_f \,, \quad \delta_{lr} = \delta_{rr} = \delta_r \tag{4.15}$$

the yaw motion can then be described by the following equation:

$$\begin{aligned} I_{zz}\ddot{r} &= M_{wz} - F_{xlf}T/2 + F_{xrf}T/2 - F_{xlr}T/2 + F_{xrr}T/2 + F_{ylf}l_f + F_{yrf}l_f - F_{ylr}l_r - F_{yrr}l_r \\ &= M_{wz} + \begin{bmatrix} -T\cos\delta_{lf}/2+l_f\sin\delta_f \\ T\cos\delta_{rf}/2+l_f\sin\delta_f \\ -T\cos\delta_{lr}/2-l_r\sin\delta_r \\ T\cos\delta_{rr}/2-l_r\sin\delta_r \end{bmatrix}^T \begin{bmatrix} F_{xwlf} \\ F_{xwrf} \\ F_{xwlr} \\ F_{xwrr} \end{bmatrix} + \begin{bmatrix} T\sin\delta_{lf}/2+l_f\sin\delta_f \\ -T\sin\delta_{rf}/2+l_f\cos\delta_f \\ T\sin\delta_{lr}/2-l_r\cos\delta_r \\ -T\sin\delta_{rr}/2-l_r\cos\delta_r \end{bmatrix}^T \begin{bmatrix} F_{ywlf} \\ F_{ywrf} \\ F_{ywlr} \\ F_{ywrr} \end{bmatrix} \end{aligned} \tag{4.16}$$

where M_{wz} is the yaw moment induced by crosswind.

The rotation for each wheel can be derived from the angular acceleration and the moment induced by the longitudinal tire force. It can be defined as

$$I_w\dot{\omega}_i = -F_{xwi}R_{wi} - f_rR_{wi}F_{zi} \,, \quad i = lf, rf, lr, rr \tag{4.17}$$

where I_w is the moment of inertia for each wheel, ω_i is the wheel angular velocity, f_r is the rolling resistance coefficient, R_{wi} is the wheel radius and F_{zi} is the normal tire force.

In Figure 4.4, the relationship between the slip angle α_i and the wheel center velocity u_{wi} is represented. The component u_{wxi} of u_{wi} in the longitudinal direction of the wheel plane can be obtained by the vehicle speed (V_x, V_y) and the yaw speed $\dot{r}$, i.e.

$$\begin{bmatrix} u_{wxlf} \\ u_{wxrf} \\ u_{wxlr} \\ u_{wxrr} \end{bmatrix} = \begin{bmatrix} 1 & 0 & 0 & 0 \\ 0 & 1 & 0 & 0 \\ 0 & 0 & 1 & 0 \\ 0 & 0 & 0 & 1 \end{bmatrix} \begin{bmatrix} \cos\delta_f \\ \cos\delta_f \\ \cos\delta_r \\ \cos\delta_r \end{bmatrix} \begin{bmatrix} V_x - T\dot{r}/2 \\ V_x + T\dot{r}/2 \\ V_x - T\dot{r}/2 \\ V_x + T\dot{r}/2 \end{bmatrix} + \begin{bmatrix} 1 & 0 & 0 & 0 \\ 0 & 1 & 0 & 0 \\ 0 & 0 & 1 & 0 \\ 0 & 0 & 0 & 1 \end{bmatrix} \begin{bmatrix} \sin\delta_f \\ \sin\delta_f \\ \sin\delta_r \\ \sin\delta_r \end{bmatrix} \begin{bmatrix} V_y + l_f\dot{r} \\ V_y + l_f\dot{r} \\ V_y - l_r\dot{r} \\ V_y - l_r\dot{r} \end{bmatrix} \tag{4.18}$$

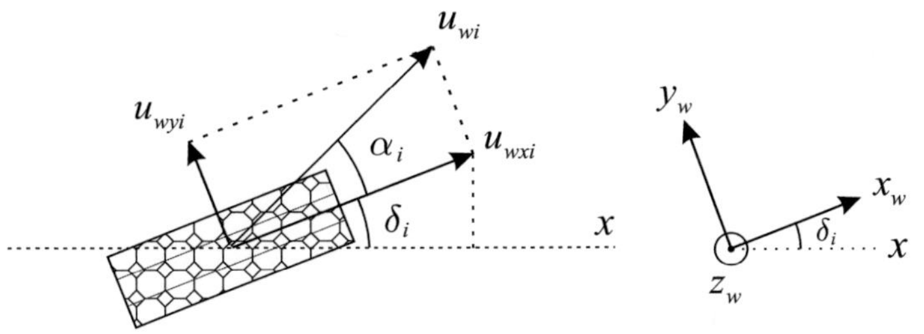

Figure 4.4: Wheel center velocity

Similarly, the component of wheel center velocity in y direction can be obtained by

$$\begin{bmatrix} u_{wylf} \\ u_{wyrf} \\ u_{wylr} \\ u_{wyrr} \end{bmatrix} = \begin{bmatrix} 1 & 0 & 0 & 0 \\ 0 & 1 & 0 & 0 \\ 0 & 0 & 1 & 0 \\ 0 & 0 & 0 & 1 \end{bmatrix} \begin{bmatrix} \sin\delta_f \\ \sin\delta_f \\ \sin\delta_r \\ \sin\delta_r \end{bmatrix} \begin{bmatrix} -V_x + T\dot{r}/2 \\ -V_x - T\dot{r}/2 \\ -V_x + T\dot{r}/2 \\ -V_x - T\dot{r}/2 \end{bmatrix} + \begin{bmatrix} 1 & 0 & 0 & 0 \\ 0 & 1 & 0 & 0 \\ 0 & 0 & 1 & 0 \\ 0 & 0 & 0 & 1 \end{bmatrix} \begin{bmatrix} \cos\delta_f \\ \cos\delta_f \\ \cos\delta_r \\ \cos\delta_r \end{bmatrix} \begin{bmatrix} V_y + l_f\dot{r} \\ V_y + l_f\dot{r} \\ V_y - l_r\dot{r} \\ V_y - l_r\dot{r} \end{bmatrix} \tag{4.19}$$

Moreover, to study the vehicle movement in the horizontal plane, the longitudinal wheel slip S_i is usually defined, which can be described by

$$S_i = \begin{cases} (u_{wxi} - \omega_i R_w)/u_i, & u_{wxi} > \omega_i R_w \\ (\omega_i R_w - u_{wxi})/u_i, & u_{wxi} < \omega_i R_w \end{cases}, \quad i = lf,\ rf,\ lr,\ rr \tag{4.20}$$

So far, a lot of tire models have been proposed to calculate the complicated and highly non-linear tire force. Smith [185] studied the tire characteristics and gave a detailed description on the parameters that have large influence on the tire force while Allen [2] discussed the modeling requirements of the tire for simulating the vehicle dynamics appropriately. More tire models can be seen in [19] [25] and [191]. Among these models, one of the most famous models is the Magic-formula model [152], which yields the tire characteristic based on a semi-empirical model with curve fitting. However, to determine the coefficients of such a model, a lot of experiments will be needed.

Unlike the Magic-formula model, the Dugoff model [28] [77], which has also been widely used in the past, needs relatively less parameters. Here, it has been successfully utilized to simulate the longitudinal and lateral tire forces. Neglecting the self-aligning moment, the tire force of Dugoff model can be determined by the following equations [28]:

$$F_{xi} = \frac{C_{si} S_i}{(1 - S_i)} f(\zeta_i) \tag{4.21}$$

$$F_{yi} = \frac{C_{\alpha i} \tan\alpha}{(1 - S_i)} f(\zeta_i) \tag{4.22}$$

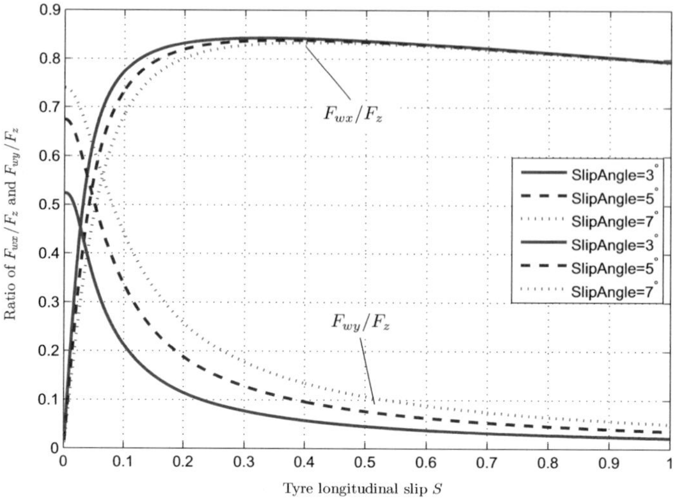

Figure 4.5: An example of tire force rations by Dugoff model

where C_{si} and $C_{\alpha i}$ are the longitudinal and lateral cornering stiffness, respectively. $f(\zeta_i)$ is a function which can be defined as follows

$$f(\zeta_i) = \begin{cases} (2-\zeta_i)\zeta_i & , \zeta_i < 1 \\ 1 & , \zeta_i \geq 1 \end{cases} \tag{4.23}$$

where

$$\zeta_i = \frac{\mu_i F_{zi}(1-S_i)\left(1-\kappa_r u_{wxi}\sqrt{S_i^2+\tan^2\alpha_i}\right)}{2\sqrt{(C_{si}S_i)^2+(C_{\alpha i}\tan\alpha_i)^2}} \tag{4.24}$$

in which α_i is the slip angle of the wheel, cf. Figure 4.4. F_{zi} is the normal tire force, μ_i is the friction coefficient of the road and κ_r is the reduction factor of the road adhesion.

Figure 4.5 shows comparison of the ratios between the longitudinal/lateral tire force and the normal tire force by Dugoff model with different slip angles. As is shown in the figure, the ratio of the longitudinal force to the normal force becomes smaller when the wheel slip angle increases. In contrast, the ratio of the lateral force to the normal force becomes larger with the increasing slip angles.

4.1.2 Driver model

As mentioned before, to be more realistic, a driver model has to be taken into account. The driver model is expected to adjust the steering angle according to the vehicle state,

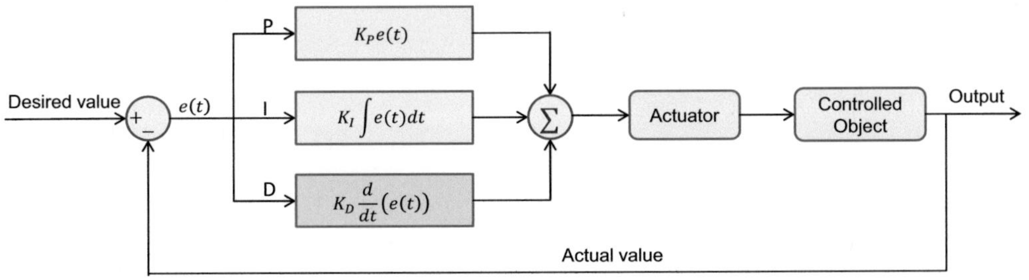

Figure 4.6: Diagram of the PID control

including the lateral displacement and yaw angle of the vehicle. Until now, a great amount of work has been carried out for modeling the driver, see [156] [97] [75] and [144].

Compared to other methods (e.g. fuzzy logic control, neutral networks and hybrid approaches) in modeling the driver, the proportional-integral-derivative (PID) method is considered to be easier to operate [141]. It has been widely used in the past (e.g. [140] and [159]) and has shown a good validity and efficiency in the practical use [165]. Owing to this, a driver model based on a PID controller is proposed and utilized in here.

Figure 4.6 illustrates a diagram of the PID control. As can be seen, the PID controller calculates an error between the measured value and the desired value. It attempts to minimize the error by an adjusting process through the actuator (here, i.e. the steering wheel). Taking the lateral displacement as well as the yaw angle and side-slip angle of the vehicle as control objectives, the corresponding PID controller for a drivel model can be described as follows:

$$\delta_f = \delta_1 + \delta_2 \tag{4.25}$$

$$\delta_1 = K_{P1}\,(Y - Y_d) + K_{I1} \int (Y - Y_d)\,\mathrm{d}t + K_{D1} \frac{\mathrm{d}}{\mathrm{d}t}\,(Y - Y_d) \tag{4.26}$$

$$\delta_2 = K_{P2}\,(\Psi - \Psi_d) + K_{I2} \int (\Psi - \Psi_d)\,\mathrm{d}t + K_{D2} \frac{\mathrm{d}}{\mathrm{d}t}\,(\Psi - \Psi_d) \tag{4.27}$$

where δ_f is the steering angle of the front wheel which is implemented here as output of the actuator in the PID system. Y and Y_d are the actual value and ideal value for the lateral displacement of the vehicle. K_{p1}, K_{I1} and K_{D1} are the corresponding control coefficients which are usually adjusted based on experience. Ψ is the sum of the yaw and side slip angle of the vehicle and Ψ_d refers to the desired value. K_{I2} and K_{D2} also refer to the corresponding control coefficients. The desired values of Y_d and Ψ_d for a vehicle running on a straight line are usually set as 0.

4.1.3 Stochastic road irregularities

Realistic roads are not exactly smooth but usually contain irregularities. Here, a general random input of the road surface with four wheels correlated (front and rear wheels are correlated with time, left and right wheels are correlated with the track width) [216] has been applied to simulate road irregularities. This stochastic road model is applicable to various kinds of vehicles and can be easily implemented.

It is expressed by the following equation

$$\dot{Z}(t) = AZ(t) + B\zeta_r(t) \tag{4.28}$$

where

$$Z(t) = [z_1(t), z_2(t), z_3(t), z_4(t), q_1, q_2]$$

$$A = \begin{bmatrix} S_1 & 0 & 0 & 0 & 0 & 0 \\ S_1 - 12V_x/T & 0 & 0 & 0 & 0 & 1 \\ -S_1 + 2V_x/L & 0 & -2V_x/L & 0 & 0 & 0 \\ -S_1 + 12V_x/T & 2V_x/L & 0 & -2V_x/L & 0 & -1 \\ -12V_x/T & 0 & 0 & 0 & 0 & 1 \\ 72V_x^2/T^2 & 0 & 0 & 0 & -12V_x^2/T^2 & -6V_x/T \end{bmatrix}$$

$$S_1 = -2\pi f_0 V_x$$

$$B = 2\pi f_0 \sqrt{V_x S_z} \cdot [\ 1 \quad 1 \quad -1 \quad -1 \quad 0 \quad 0\ ]^T$$

V_x is the longitudinal vehicle speed. $L = l_f + l_r$ is the wheelbase and T refers to the vehicle track, $\zeta_r(t)$ is a white noise, z_i, $i = 1, 2, \cdots, 4$ refers to the road irregularity for four wheels, respectively. q_1, q_2 are the intermediate variables, f_0 is the cut-off frequency of the road surface and S_z is the road roughness coefficient.

An example of the road input to four wheels based on the above model can be seen in Figure 4.7.

4.1.4 Vehicle response under crosswind excitation

To study the dynamic response of the vehicle under strong crosswind, a series of simulations have been carried out. The wind model, road model and driver model are combined with the vehicle model in MATLAB/Simulink. A sketch of the simulation diagram can be seen in Figure 4.8.

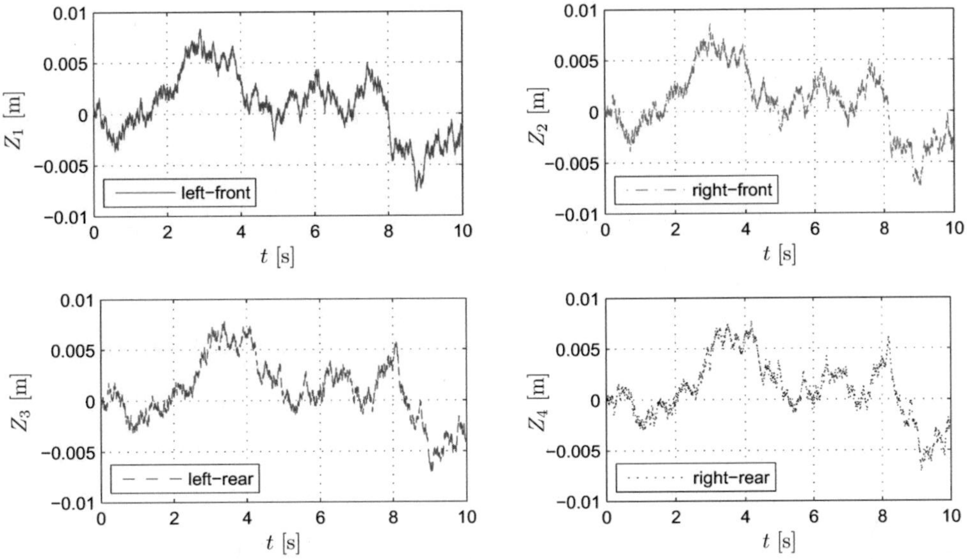

Figure 4.7: Simulation results of road irregularities for four wheels

The main parameters that characterize the truck model are listed in Table 4.1. For parameters of other vehicles please refer to the appendix. These parameters are required for the simulation and chosen to be representative of a range of vehicles within each vehicle category. The geometric dimensions (including vehicle lengths, vehicle heights and the distances between c.g. and front/rear axle) as well as the masses of the vehicles are based on experience. Other parameters (e.g. the suspension parameters and the corresponding moment of inertia) of each vehicle type are taken from published papers, cf. truck [45] [9], bus [113] [45] [9], sprinter [45] [9] [28], trailer [208] [166] [28] and car [28] [9]. Due to the lack of reference values, the yawing moment of inertia of the light trailer has been calculated approximately based on a rectangular vehicle shape. For tires, the corning stiffness depends on several parameters, such as tire size, wheel width, tire materials and so on. For a given tire, it depends on the vertical load and inflation pressure. Therefore, it is difficult to determine an accurate value of the cornering stiffness for each tire. Without loss of generality, in this work, an example value of the tire cornering stiffness from [110] has been applied for the truck model (for other vehicles please refer to [110] [28] [208] and [210]). Besides, it should be mentioned again that the vehicle masses and moments of inertia here are considered for unladen vehicles.

In addition, to follow a target route, it is necessary for the driver to minimize the errors between the vehicle actual response and the desired response. This can be realized by adjusting the control parameters (K_p, K_I and K_D) of the PID controller. In reality, each driver may react differently. Therefore, in the presented model, the control parameters of

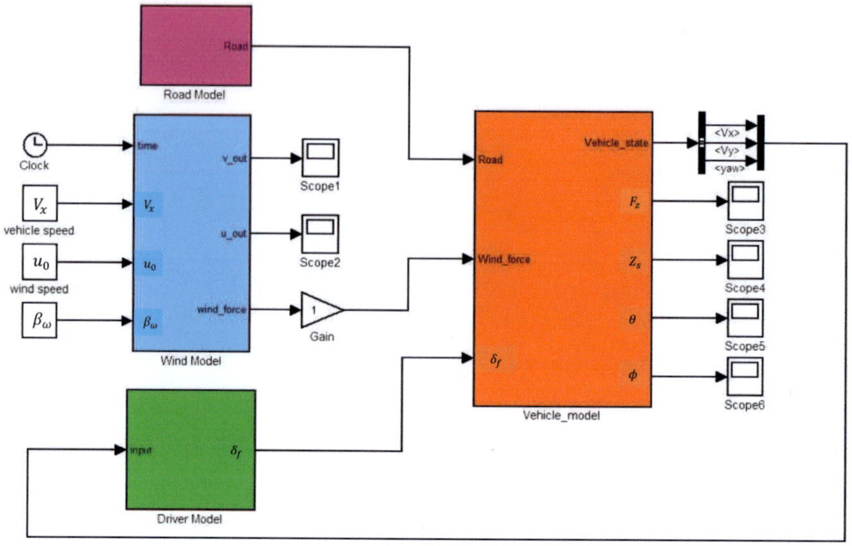

Figure 4.8: Simulation diagram for the vehicle-wind system

the PID controller are determined based on the fact that the driver model should yield a reasonable result, i.e. the control objectives (e.g. the lateral displacement and the yaw angle) oscillate but tend to a limited small value under certain wind excitations, cf. [10] [4]. In practice, one method for determining the control parameter is to first set K_I and K_D to zero and then increase K_P until the output oscillates. Next, set K_P to about half the value and increase K_I and K_D until the output can be accepted. The effects of increasing each parameter on the response of the system can be seen in [218] and finally the parameters of K_{P1}, K_{I1}, K_{D1} and K_{P2}, K_{I2}, K_{D2} used here are set as $0.6, 0.3, 0.3$ and $0.6, 0.8, 4$, respectively. More contents about tuning rules of the PID controller can be seen in [151] and [3], etc.

Based on the model described above, vehicle dynamic response for a truck running under crosswind is studied. The wind speed is assumed to be 0 at initial time and increases gradually until a given value u_0 in 1s and then keeps constant. An example of the artificial wind at $u_0 = 15$ m/s can be seen in Figure 4.9. The aerodynamic coefficients are taken from [186] and the vehicle speed is set as 60 km/h.

Figure 4.10 shows the lateral displacement of the vehicle under the excitation of different wind speeds. As can be seen, when the vehicle enters the crosswind, its lateral displacement increases gradually first and then varies around zero due to the control by the driver. For a low wind speed, e.g. $u_0 = 15$ m/s, the vehicle becomes stable in a short time and can be

Table 4.1: Main parameters for the truck used in the numerical simulation [9] [32] [166]

m	Vehicle mass	7500 kg
m_s	Sprung mass	6000 kg
l_f	Distance between c.g. and the front axle	1.8 m
l_r	Distance between c.g. and the rear axle	3.2 m
L	Vehicle length	7 m
H	Vehicle height	4 m
I_{xx}	Rolling moment of inertia	1352 kgm^2
I_{yy}	Pitching moment of inertia	5483 kgm^2
I_{zz}	Yawing moment of inertia	50000 kgm^2
K_{sf}	Upper vertical spring stiffness	400000 N/m
K_{sr}	Lower vertical spring stiffness	400000 N/m
C_f	Upper vertical damping stiffness	20000 Ns/m
C_r	Lower vertical damping stiffness	20000 Ns/m
$C_{\alpha f}$	Cornering stiffness of one tire (front)	65682 N/rad
$C_{\alpha r}$	Cornering stiffness of one tire (rear)	131365 N/rad

well controlled. With the increase of the wind speed, the corresponding lateral acceleration and yaw angle become larger and the time for the vehicle back to a stable state turns to be longer (cf. Figure 4.11 and Figure 4.12), which implies that higher crosswind speed will make it more difficult for the driver to control the vehicle. As can be imagined, when the wind speed is large enough, the vehicle may lose control, which will affect the traffic or even cause accidents.

Figure 4.13 represents the wheel normal forces of the vehicle under different wind speeds. The wheel normal forces of the leeward (left) and windward (right) wheels at the same axle are equivalent at the very beginning. However, under crosswind, this balance will be changed. Normal forces of the leeward wheels become larger due to the crosswind excitation. By contrast, normal forces of the windward wheels turn to be smaller. It is foreseeable that the wheel normal force of the right-rear wheel in the windward side will first become zero when the wind speed is extreme high. This may lead to a risk of overturning for the vehicle.

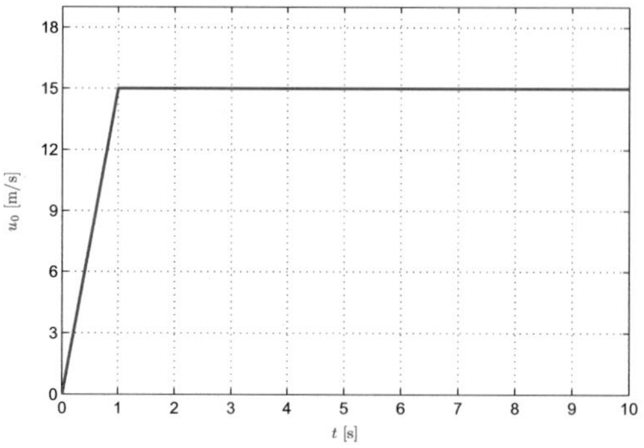

Figure 4.9: Crosswind model with constant wind speed $u_0 = 15$ m/s

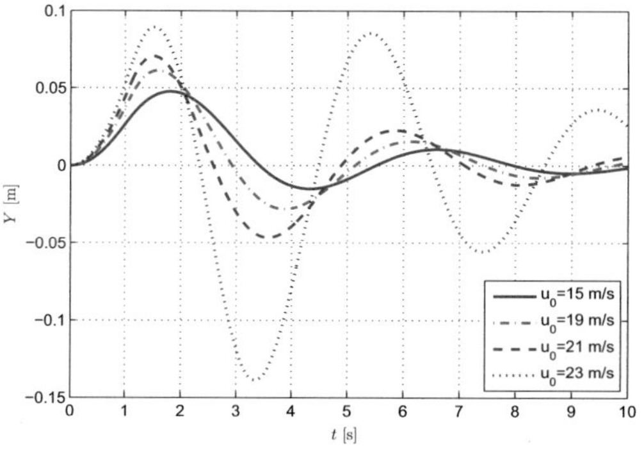

Figure 4.10: Lateral displacement of the vehicle for different wind speeds

Besides the wind speed and vehicle speed, the driver's behavior has also an influence on crosswind stability. Figure 4.15 shows the steering angle of the vehicle operated by a driver under different wind speeds. As is shown, higher wind speeds give rise to larger steering angles. At a higher wind speed, more time will be needed for the driver to adjust the vehicle back to a stable state. In Figure 4.16, the lateral displacement of the vehicle controlled by two different driver models are represented. The different driver models are modeled by setting different parameters of the PID controller. Here, the parameters of PID controller K_{P1}, K_{I1}, K_{D1} are set as $0.6, 0.3, 0.3$ and $0.6, 0.3, 0.2$, respectively. As can be seen in the figure, the first driver can control the vehicle much better than the second driver at the same wind speed.

Figure 4.14 depicts the influence of road irregularities on the wheel force (here, take the right-front wheel for example). It can be seen that the road irregularities have an effect on the wheel force.

Moreover, Figure 4.18 compares the lateral displacement of the vehicle running in crosswind with or without gust, cf. Figure 4.17. The mean gust wind speed u_0 is set as 15 m/s while

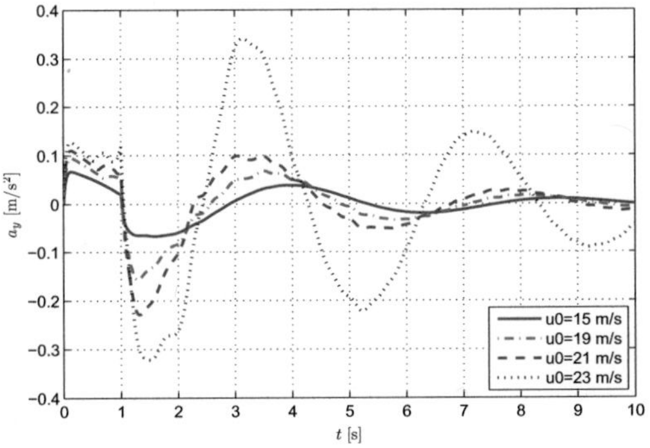

Figure 4.11: Lateral acceleration of vehicle for different wind speeds

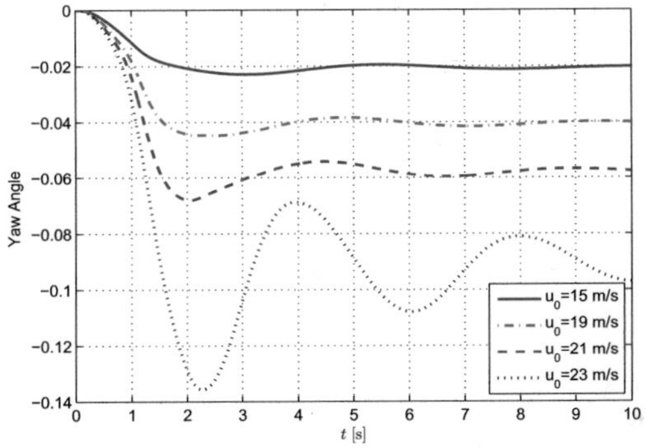

Figure 4.12: Yaw angle of vehicle with different wind speeds

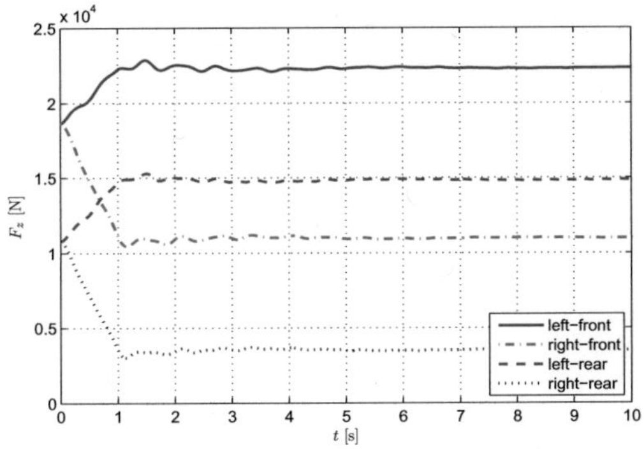

Figure 4.13: Wheel normal forces of the vehicle under crosswind excitation at $u_0 = 15$ m/s

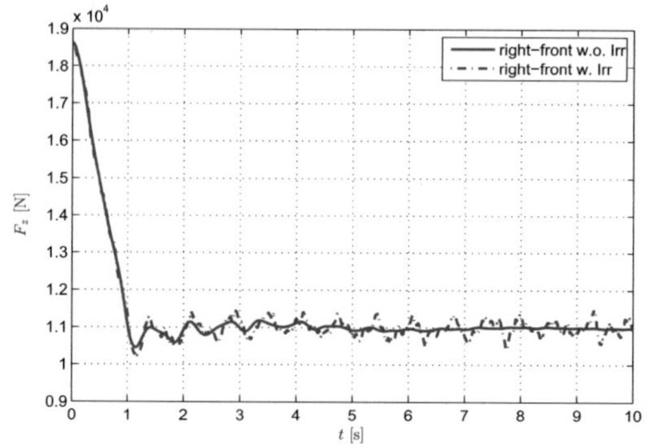

Figure 4.14: Influence of road irregularities on the wheel force

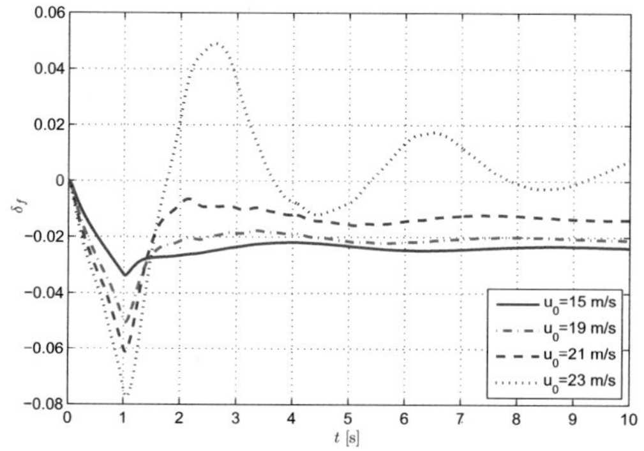

Figure 4.15: Steering angle of the vehicle for different wind speeds

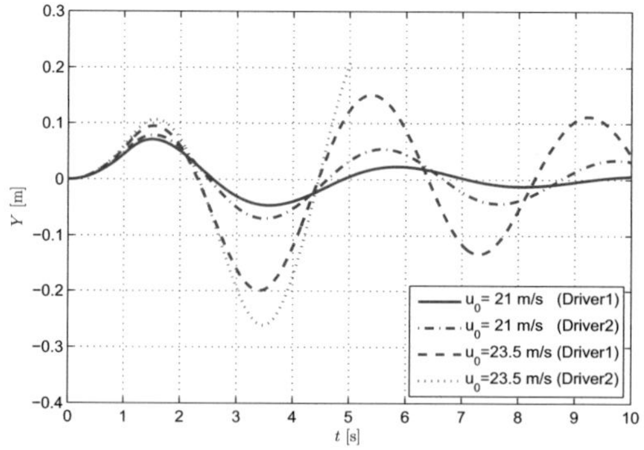

Figure 4.16: Influence of the driver on lateral displacements

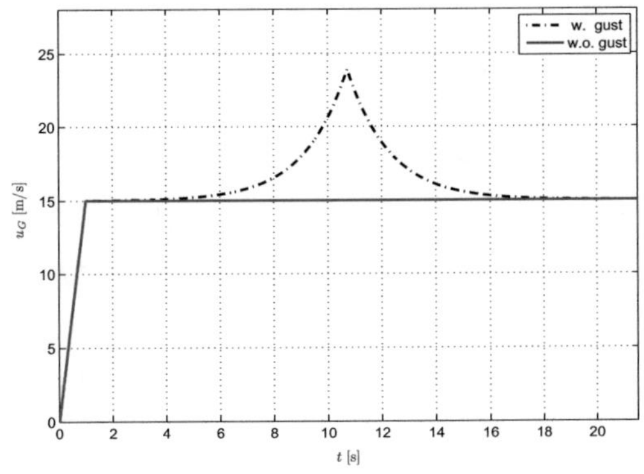

Figure 4.17: Crosswind model with or without gust

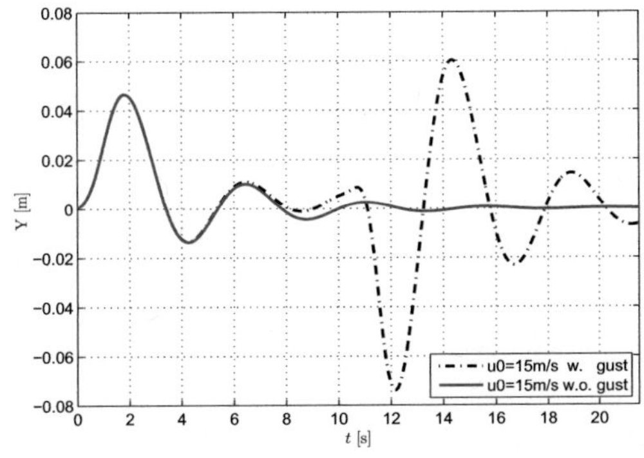

Figure 4.18: The lateral displacement of the vehicle under crosswind excitation with or without gust

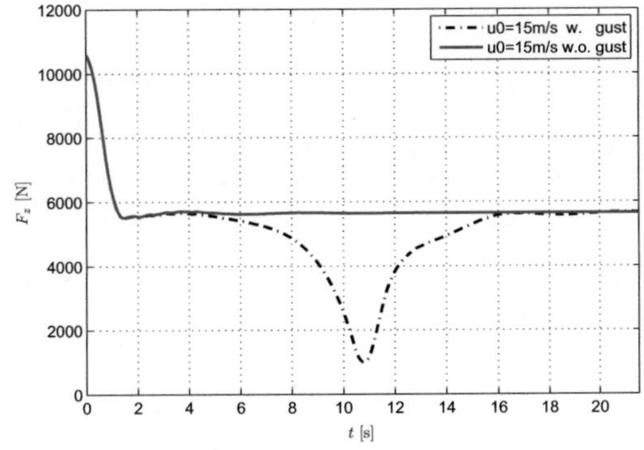

Figure 4.19: The normal wheel force of the vehicle in the crosswind with or without gust (right-rear wheel)

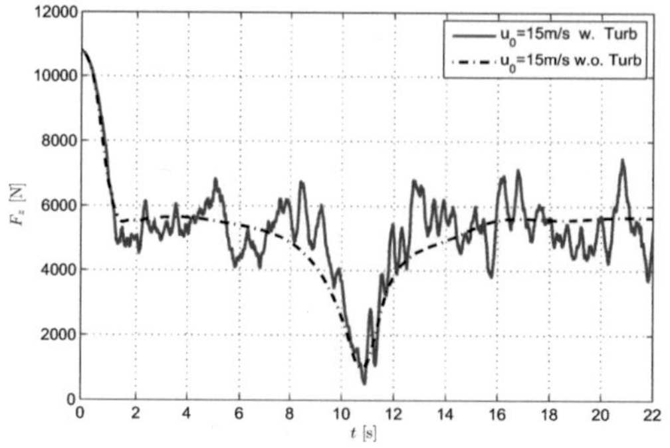

Figure 4.20: Influence of the wind turbulence on the wheel normal force (right-rear wheel)

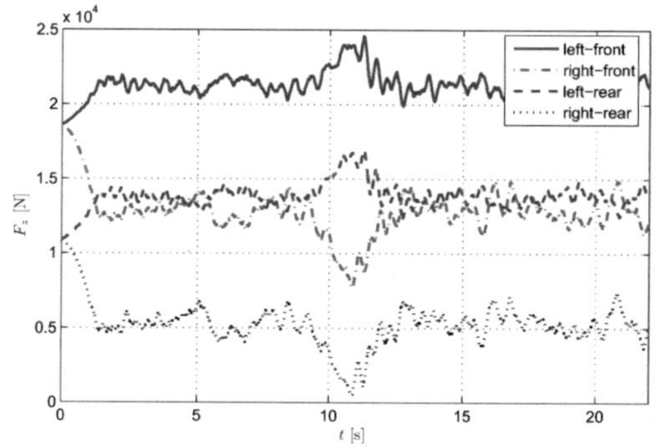

Figure 4.21: Normal wheel forces of the vehicle under the excitation of gust with turbulence

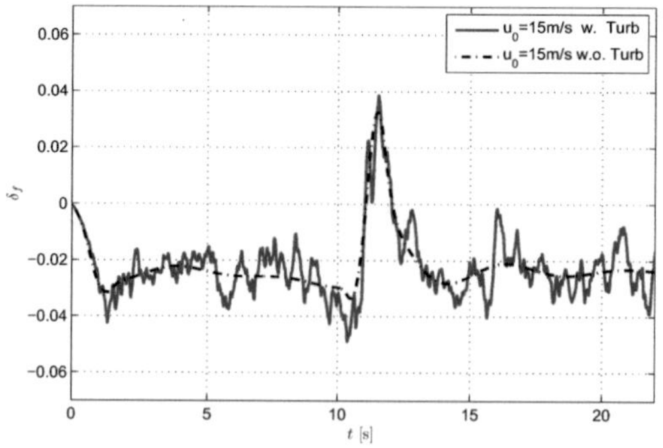

Figure 4.22: Steering angle of the vehicle running in the gust with or without turbulence

the mean vehicle speed is set as 60 km/h. As can be seen from the figure, a deviation of the vehicle under the excitation of a gust is much larger than that under the excitation of wind without gust. For the former, the largest deviation happens at the time when the gust peak arises. In comparison, the vehicle running in the crosswind without gust is more stable.

Corresponding wheel normal forces of the vehicle (take the rear wheel on the leeward side, i.e. the right-rear wheel for example) excited by the wind with or without gust can be seen in Figure 4.19. As is shown, the normal wheel force decreases largely when the gust appears.

Figure 4.20 represents the response of the right-rear wheel of the vehicle running in the gust with or without turbulence. It can be seen that when the turbulence is taken into account, the wheel normal forces change more frequently. Figure 4.21 shows the corresponding normal forces of all the wheels, in which the influence of wind turbulence on each wheel can be well seen. In Figure 4.22, a comparison of the steering angle in the gust with or without turbulence has been represented. It can be concluded that the driver's action in a gust scenario with turbulence occurs more frequently. This may cause the fatigue of the driver and thereby affect the safety of the vehicle.

4.1.5 Car-trailer model

Nowadays, coupled vehicles, which consist of the powered and towed parts, e.g. a passenger car with trailer, are becoming more and more widely used in road transport [208]. The driving behavior of coupled vehicles is significantly different from that of the non-coupled vehicles [106]. To study the dynamic performance of coupled vehicles, a simplified linear single-track model [93] [192] is usually applied, cf. Figure 4.23, in which the tires at each axle are represented as a single tire and the steering angle as well as the hitch angle between the car and the trailer are usually considered to be small.

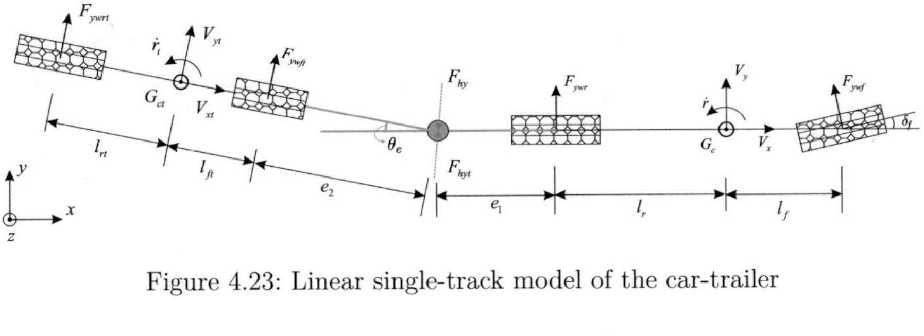

Figure 4.23: Linear single-track model of the car-trailer

The linear model is easy to be implemented for the analysis of the lateral stability of the car with trailer. However, when the crosswind is considered, the linear model will not be sufficiently accurate anymore due to the fact that the hitch angle cannot be considered as small variable again and both the lateral and longitudinal performance of the vehicle

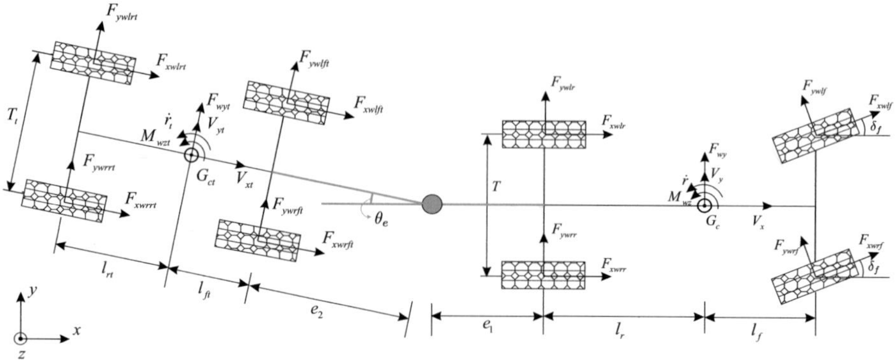

Figure 4.24: Nonlinear double-track model of the car-trailer in yaw plane

should be taken into account. In other words, to analyze the crosswind stability of car-trailers, a nonlinear car-trailer model has to be established. In this work, the nonlinear car-trailer model is developed based on a double-track car-trailer model as can be seen in [110] and [106]. Other parts like the tire and road models are built the same as before.

Figure 4.24 shows the nonlinear double-track model for a car-trailer in the yaw plane. When the nonlinear model is considered, the steering angle as well as the hitch angle will not be regarded as small variables any more. Mathematical analysis for the car-trailer model can be carried out based on the two separate parts, i.e. the car and the trailer, respectively. The whole car-trailer system can be combined by a hitch force according to its physical model.

Main equations of motion for the nonlinear car-trailer model can be listed as follows

$$m(\dot{V}_x - \dot{r}\cdot V_y) = (F_{xwlf}+F_{xwrf})\cdot\cos(\delta_f) - (F_{ywlf}+F_{ywrf})\cdot\sin(\delta_f) + F_{xwlr} + F_{xwrr} + F_{hx} \quad (4.29)$$

$$m(\dot{V}_y + \dot{r}\cdot V_x) = (F_{xwlf}+F_{xwrf})\cdot\sin(\delta_f) + (F_{xwlf}+F_{xwrf})\cdot\sin(\delta_f) + F_{ywlr} + F_{ywrr} + F_{hy} + F_{wy} \quad (4.30)$$

$$\begin{aligned} I_{zz}\cdot\ddot{r} = &-(F_{xwlf}\cdot\cos(\delta_f) - F_{ywlf}\cdot\sin(\delta_f))\cdot\tfrac{T}{2} + (F_{xwrf}\cdot\cos(\delta_f) - F_{ywrf}\cdot\sin(\delta_f))\cdot\tfrac{T}{2} \\ &+ F_{xwlr}\cdot\tfrac{T}{2} - F_{xwrr}\cdot\tfrac{T}{2} + (F_{xwlf}\cdot\sin(\delta_f) + F_{ywlf}\cdot\cos(\delta_f))\cdot l_f + (F_{xwrf}\cdot\sin(\delta_f) \\ &+ F_{ywrf}\cdot\cos(\delta_f))\cdot l_f - (F_{ywlr} + F_{ywrr})\cdot l_r - F_{hy}\cdot(l_r + e_1) + M_{wz} \end{aligned} \quad (4.31)$$

$$m_t(\dot{V}_{xt} - \dot{r}_t\cdot V_{yt}) = F_{xwlft} + F_{xwrft} + F_{xwlrt} + F_{xwrrt} + F_{hxt} \quad (4.32)$$

$$m_t(\dot{V}_{yt} + \dot{r}_t\cdot V_{xt}) = F_{ywlft} + F_{ywrft} + F_{ywlrt} + F_{ywrrt} + F_{hyt} + F_{wyt} \quad (4.33)$$

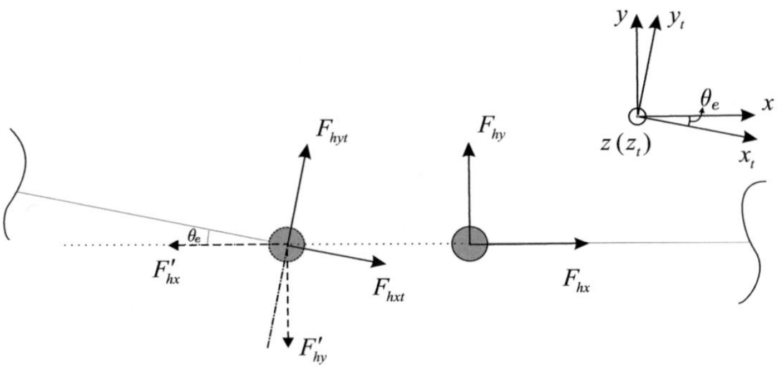

Figure 4.25: Hitch force between the car and trailer

$$\begin{aligned} I_{zzt}\cdot\ddot{r}_t = &\left(-F_{xwlft}+F_{xwrft}-F_{xwlrt}+F_{xwrrt}\right)\cdot\tfrac{T}{2}+\left(F_{ywlft}+F_{ywrft}\right)\cdot l_{ft} \\ &-\left(F_{ywlrt}+F_{ywrrt}\right)\cdot l_{rt}-F_{hyt}\cdot\left(l_{ft}+e_2\right)+M_{wzt} \end{aligned} \tag{4.34}$$

where the subscript t means the towed part, i.e. the trailer. e_1 is the distance between the hitch point and rear axle of the car while e_2 is the distance from the hitch point to the front axle of the trailer. F_{hx}, F_{hy}, F_{hxt} and F_{hyt} represent the hitch force acting on the car and trailer in the longitudinal and lateral direction, respectively. F_{wy} and F_{wyt} are corresponding wind forces in the lateral direction.

Figure 4.25 illustrates the relationship of the hitch force between the car and the trailer. F'_{hx} and F'_{hy} refer to the reaction forces of F_{hx} and F_{hy}. The corresponding equation can be expressed as follows

$$F_{hxt} = F_{hy}\cdot\sin(\theta_e) - F_{hx}\cdot\cos(\theta_e) \tag{4.35}$$

$$F_{hyt} = -F_{hy}\cdot\cos(\theta_e) - F_{hx}\cdot\sin(\theta_e) \tag{4.36}$$

where θ_e is the hitch angle between the car and trailer.

4.2 Railway vehicles

Unlike the modeling for road vehicles, the railway vehicles, including tracks, are modeled in the commercial MBS software ADAMS/Rail. Based on ADAMS/Rail, a classical double-decker cabin car model [202] has been established. It contains nonlinear spring and damper characteristics. In addition, bump stops that have great influence on the overturning behavior of the vehicle have been also taken into consideration. The wheel-rail forces are calculated by Kalker's FASTSIM algorithm, which is considered to be a good compromise between the simulation speed and simulation accuracy [115].

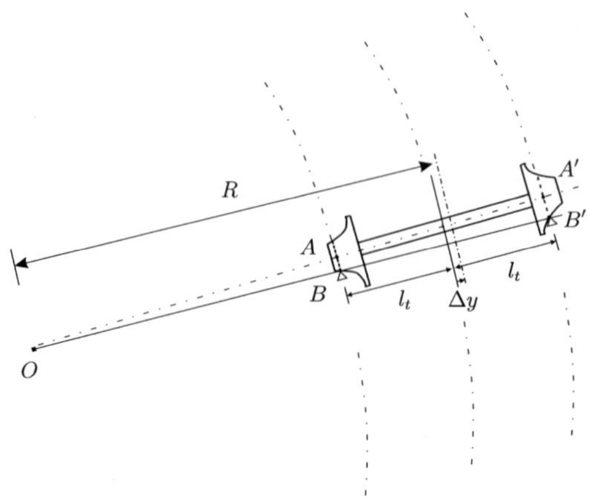

Figure 4.26: The wheelset with coned wheels rolls on a curved track [206]

Before studying the crosswind stability of railway vehicles running on straight/curved track, parameters regarding the track geometry and corresponding wind scenarios for the vehicles running on curved track are introduced first.

According to Redtenbacher's theory, a wheel set with coned wheels of the railway vehicle running on curved track can maintain a pure rolling motion if it moves outward and adopts a radial position [206]. This can be illustrated in Figure 4.26, the corresponding geometric conditions can be obtained by

$$\frac{r_0 - \lambda \Delta y}{r_0 + \lambda \Delta y} = \frac{R - l_t}{R + l_t} \tag{4.37}$$

which, for simplicity, is usually simplified as

$$\Delta y = \frac{r_0 l_t}{\lambda_0 R} \tag{4.38}$$

where r_0 and λ_0 are the nominal rolling radius and wheel conicity, respectively. l_t is half of the track distance which is measured on the top track plane ($2l_t = 1500$ mm for the standard track). R is the horizontal curve radius.

This theoretical analysis neglects friction forces on the wheel set and represents a good simplification. However, in reality, the pure rolling seldom happens for various complicated reasons (e.g. wheel and track geometry errors may cause instability to the vehicle). Lateral acceleration occurs when the vehicle runs on curved tracks. To compensate part of the lateral acceleration, the track cant is usually designed for the construction of railway tracks. It is defined as the difference between the levels of two rails, i.e. the inner and outer rail,

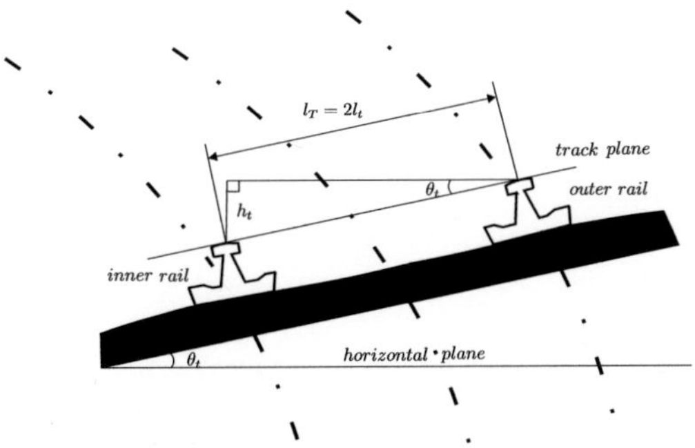

Figure 4.27: The definition of cant for a curved track [134]

cf. Figure 4.27. Referred to a horizontal plane, the outside rail rises above the inner rail with a cant angle θ_t, which is defined by

$$\theta_t = \arcsin\left(\frac{h_t}{l_T}\right) = \arcsin\left(\frac{h_t}{2l_t}\right) \tag{4.39}$$

where h_t is the track cant, l_T is the nominal distance between the wheel-rail contact points of a wheelset.

Due to the existence of the track cant, attention has to be paid that when studying the crosswind stability of the vehicle running on curved track, the wind directions (from inner side or outer side along the vehicle running orientation) have to be identified. This is different from the case of the vehicle running on straight track.

Figure 4.28 illustrates two running scenarios for the vehicle passing through the curved track with inner- or outer- wind excitation. When the vehicle speed becomes high, the component of vehicle gravity in the track plane may not be enough to support the centripetal force any more. For the outer-side wind excitation scenario (Figure 4.28(a)), the wind force will first counteract the centrifugal force and then if it is strong enough, the vehicle will have a risk of overturning. For the other scenario, i.e. the wind comes from the inner side (cf. Figure 4.28(b)), the crosswind acting in the same direction with the centrifugal force will make the vehicle be more prone to be blown over.

Figure 4.29 shows the lateral acceleration a_y and resultant acceleration angle ς in the track plane for the inner-side wind excitation scenario. According to the figure, the corresponding quasi-static lateral acceleration in the track plane can be obtained

$$a_y = \frac{v_0^2}{R}\cos\theta_t - g\sin\theta_t = \frac{v_0^2}{R}\cos\theta_t - g\frac{h_t}{l_T} \tag{4.40}$$

Corresponding vertical acceleration is

$$a_z = \frac{v_0^2}{R}\cdot\frac{h_t}{l_T} + g\cos\theta_t \tag{4.41}$$

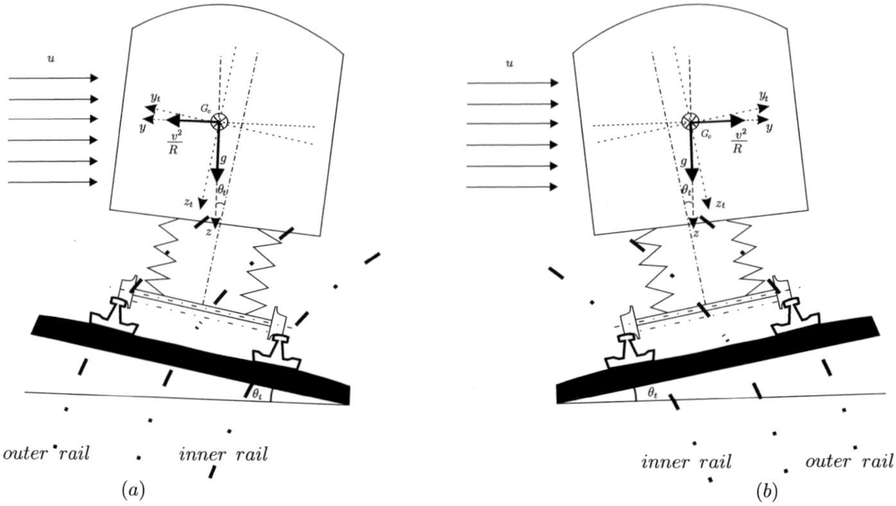

Figure 4.28: Wind directions for the vehicle running on a curved track ((*a*).outer side and (*b*). inner side)

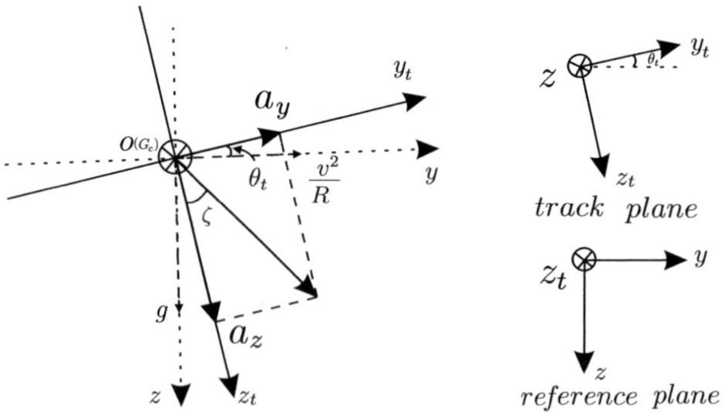

Figure 4.29: The lateral acceleration a_y and lateral force angle ς in the track plane(inner side)

For small cant angle ($\theta_t \approx 0$), Equation (4.40) can be approximated as

$$a_y \approx \frac{v_0^2}{R} - g\frac{h_t}{l_T} \tag{4.42}$$

Thus, the equilibrium cant which gives $a_y = 0$ for a given vehicle speed and a given radius can then be obtained by

$$h_{eq} = \frac{l_T}{g} \cdot \frac{v_0^2}{R} \tag{4.43}$$

In reality, due to variations of vehicle speed (e.g. vehicle running with braking or at a much slower/quicker speed or different vehicles running in the same track, etc.), the actual cant h_t is usually less than the equilibrium cant h_{eq}. A difference between h_{eq} and h_t is called cant deficiency which can be given by

$$h_e = h_{eq} - h_t = \frac{l_T}{g} \cdot \frac{v_0^2}{R} - h_t \tag{4.44}$$

Combining Equations (4.42), (4.43) and (4.44), one obtains

$$h_t = \frac{l_T}{g} \cdot \frac{v_0^2}{R} - \frac{l_T}{g} \cdot a_y \tag{4.45}$$

For a given horizontal curve radius R and track cant h_t, the permissible vehicle speed v_p can be defined by the permissible cant deficiency $h_{e,p}$ or the permissible lateral acceleration $a_{y,p}$ by Equation (4.46) [134]:

$$v_p = \sqrt{\frac{gR}{2l_t}(h_{e,p} + h_t)} = \sqrt{R(a_{y,p} + g \cdot \frac{h_t}{2l_t})} \tag{4.46}$$

Figure 4.30 shows an example of the permissible vehicle speed v_p with relation to the permissible cant deficiency $h_{e,p}$ and lateral acceleration $a_{y,p}$ on a standard gauge track with $R = 3200$ m and $h_t = 100$ mm. As can be expected, with the increase of $h_{e,p}$ and $a_{y,p}$, the permissible vehicle speed v_p becomes larger.

In addition, it should be pointed out that in the construction of railway tracks, in order to make the track curvature and the lateral acceleration of the vehicle change gradually, a transition track is usually built between the tangent track and the circular track or between two adjacent curved tracks. The transition track with a linear variation of the curvature

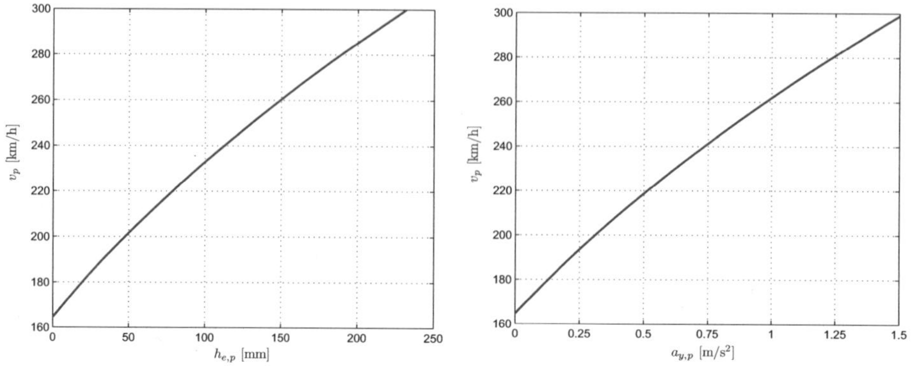

(a) v_p vs. $h_{e,p}$ (b) v_p vs. $a_{y,p}$

Figure 4.30: An example of the permissible vehicle speed v_p with relation to $h_{e,p}$ and $a_{y,p}$ on a standard gauge track with $R = 3200$ m and $h_t = 100$ mm

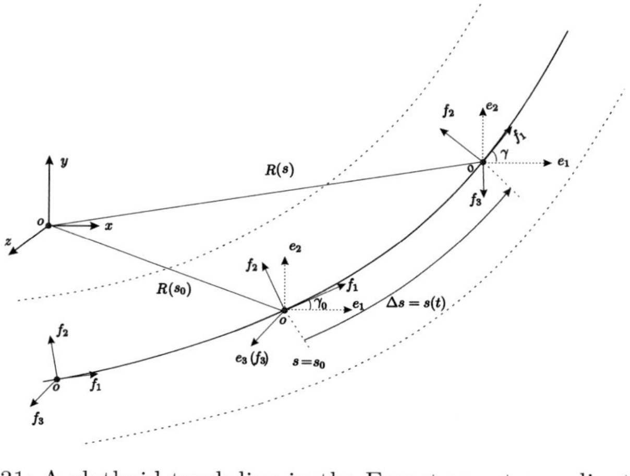

Figure 4.31: A clothoid-track line in the Frenet-serret coordinate system

is called clothoid-track. To get a kinematic model of the vehicle running along the track as well as a geometry description of the track, the Frenet-Serret coordinate system (or TNB frame) is usually applied. Figure 4.31 shows a sketch of the clothoid-track line in the Frenet-Serret coordinate system.

Based on the Frenet-Serret coordinate system, a clothoid-track line in the global coordinate system can be expressed by the following equation [85]:

$$\vec{R}(t) = \begin{pmatrix} X_0 \\ Y_0 \end{pmatrix} + \begin{pmatrix} \int_{s_0}^{s_1} \cos(\gamma(s)) d_s \\ \int_{s_0}^{s_1} \sin(\gamma(s)) d_s \end{pmatrix} = \begin{pmatrix} X_0 \\ Y_0 \end{pmatrix} + \begin{pmatrix} \int_{s_0}^{s_1} \cos(\gamma_0 + \kappa_0 s(t) + \frac{1}{2A^2} s(t)^2) d_s \\ \int_{s_0}^{s_1} \sin(\gamma_0 + \kappa_0 s(t) + \frac{1}{2A^2} s(t)^2) d_s \end{pmatrix} \tag{4.47}$$

where X_0, Y_0 are the initial coordinates of the track center, γ_0 is the azimuth of unity tangential vector at time $t = 0$. $s(t)$ refers a function of time t which indicates the arc length and s_0, s_1 are the arc length at time $t = 0, t = t_1$, respectively. κ_0 is initial curvature of the track, A is the clothoid parameter. Circle track with fixed curvature ($\kappa = \kappa_c$) and straight track without curvature ($\kappa = 0$) are the special cases of clothoid track. More details on the geometrical features of curved tracks can be found in [134] [85].

5 Risk analysis for the crosswind stability of vehicles

This chapter presents a probabilistic method for assessing the crosswind stability of vehicles running in strong crosswind. A brief introduction to the probabilistic method will be presented in the following section. Then, in the second section, failure criteria and corresponding limit-state functions for railway and road vehicles are stated. Next, the failure probabilities with respect to a single gust event of vehicles running on straight track/road or curved track are defined and methods for computation of failure probabilities are introduced. Finally, a study will be carried out for the determination of failure probabilities of a vehicle running for a relatively long time interval.

5.1 Probabilistic method

As mentioned before, simulation methods for crosswind stability of vehicles can be broadly classified into two types, namely deterministic methods and probabilistic methods.

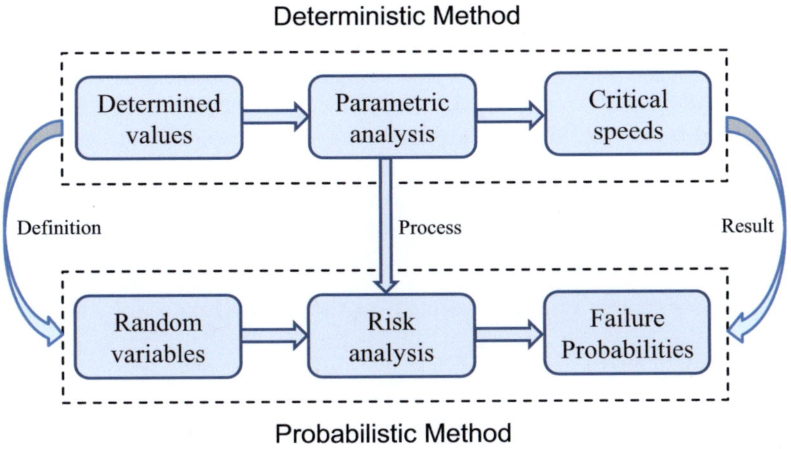

Figure 5.1: Comparison of the deterministic method and the probabilistic method for studying the crosswind stability of vehicles

Deterministic methods are characterized by critical wind speeds, at which an accident may occur for the vehicle with certain vehicle speeds. They consider all the variables as deterministic values and calculate the critical wind speed with respect to a given gust shape. On the other hand, probabilistic methods take random variables rather than determined values into account. As a result, it is no longer possible to determine the exact critical speeds for the vehicle in a certain wind scenario, but it is possible to compute the failure probabilities of the vehicle based on risk analysis.

This work studies the crosswind stability of vehicles based on a probabilistic method. A clear comparison of the deterministic method and the probabilistic method can be seen in Figure 5.1.

5.2 Definition of failure criteria

In order to assess the risk of a vehicle running under strong crosswind, failure criteria for different vehicles have to be set first so that failure events can be captured. For railway vehicles, the main failure mode due to crosswind is overturning. It is usually described by a critical wheel unloading as follows [160]

$$\delta Q = \frac{Q_{sta} - Q_{dyn}}{Q_{sta}} = 1 - \frac{Q_{dyn}}{Q_{sta}} \tag{5.1}$$

where Q_{dyn} and Q_{sta} are the dynamic and static wheel unloading on the windward side, respectively. δQ refers to the wheel unloading change. For a safety margin of 10% [70], δQ will be set to 0.9. A sketch for a railway vehicle under strong crosswind excitations can be seen in Figure 5.2.

Failure criteria for road vehicles should also consider the potential effect that the vehicle could bring to traffic. Take a sideslip for example, if the sideslip is too large, the vehicle may slide to other lanes. In this case, although the vehicle itself is not overturning, it may impact other vehicles or move to the curbs, and thus lead to accidents. According to [10] [90], there are mainly three failure modes for road vehicles occurring in strong crosswind, namely excessive sideslip, overturning and excessive rotation. They can be defined as follows:

1) sideslip accidents: the lateral displacement of the vehicle exceeds 0.5 m, i.e.

$$y_0(t) \geq y_{0\,\max} = 0.5 \text{ m} \tag{5.2}$$

where $y_0(t)$ is the actual displacement of the vehicle in lateral direction due to the crosswind and $y_{0\,\max}$ is the critical value for y_0.

2) overturning accidents: any one of the wheel unloadings in the windward side falls to zero, i.e.

$$Q_r = \min\left\{\frac{Q_{dyn\_fw}}{Q_{sta\_fw}}, \frac{Q_{dyn\_rw}}{Q_{sta\_rw}}\right\} \leq 0 \tag{5.3}$$

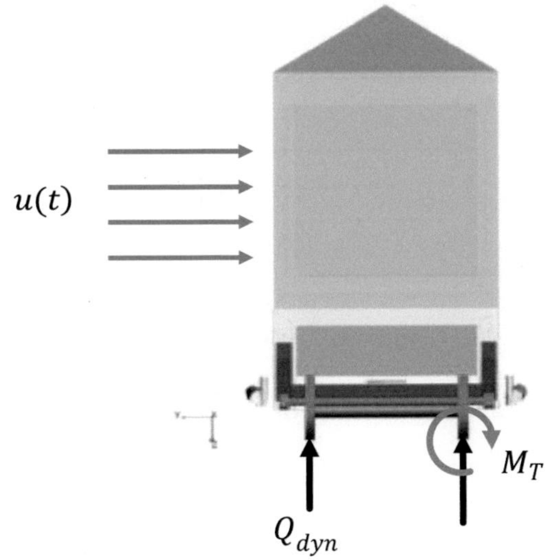

Figure 5.2: A sketch for the railway vehicle under strong crosswind excitations

where Q_r is the sign for the change of wheel force, Q_{dyn_fw}, Q_{dyn_rw} refer to the vertical force of the front wheel and rear wheel in the windward side, respectively. Q_{sta_fw} and Q_{sta_rw} are the corresponding static wheel forces.

3) rotational accidents: the rotational angle of the vehicle in the yaw plane exceeds 0.2 rad, i.e.

$$r(t) \geq r_{0\,\max} = 0.2 \tag{5.4}$$

where $r(t)$ is the actual yaw angle, $r_{0\max}$ is the critical value of $r(t)$.

Based on the above failure criteria, the failure probability for each vehicle in various different wind scenarios can be defined (see the following section). For road vehicles, an accident will be said to occur if any one of the three failure modes happens.

5.3 Reliability analysis

The first step of a reliability analysis is to map random variables on the dimensionless standard normal space $\aleph(0, 1)$ by a nonlinear transformation, e.g. Nataf transformation [135] or Rosenblatt transformation [169]. This step is a must for semi-analytical procedures and also recommended for Monte-Carlo methods [205].

Without considering the nonstationary wind turbulence, in this work, there are totally 8 random variables, i.e. the nonstationary aerodynamic coefficients C_x, C_y, C_z, C_{Mx}, C_{My}, C_{Mz}, the gust amplitude A and the gust duration T. If the nonstationary wind turbulence is also considered, the number of random variables will increase drastically to about several hundreds. In the following part, all these random variables are grouped together in a vector $\hat{\mathbf{z}}$.

Then, in the next step, a performance function $g(\hat{\mathbf{z}})$ which separates the failure domain $\mathbf{F} : g(\hat{\mathbf{z}}) \leq 0$ from the safe domain $\mathbf{S} : g(\hat{\mathbf{z}}) > 0$ has to be introduced. $g(\hat{\mathbf{z}}) = 0$ is the limit state function which refers to a critical case.

The performance function for railway vehicles in relation to the wheel unloading can be described as follows

$$g(\hat{\mathbf{z}}) = \min_{t \in [0, T_0]} \left\{ \frac{Q_{dyn}}{Q_{sta}} - 0.1 \right\} \tag{5.5}$$

For road vehicles, the corresponding performance function can be expressed by

$$g(\hat{\mathbf{z}}) = \min \left\{ y_{0\,\max} - |y_0(t)| \,,\; Q_r,\; r_{0\,\max} - |r(t)| \right\} \tag{5.6}$$

i.e.

$$g(\hat{\mathbf{z}}) = \min \left\{ 0.5 - |y_0(t)| \,,\;\; \min \left\{ \frac{Q_{dyn\_fw}}{Q_{sta\_fw}}, \frac{Q_{dyn\_rw}}{Q_{sta\_rw}} \right\},\;\; 0.2 - |r(t)| \right\} \tag{5.7}$$

Based on the performance function, the failure probability of the vehicle for a single gust event can be defined as follows

$$P_G = P_f(u_0, v_0, \alpha_\omega) = \int\limits_{g(\hat{\mathbf{z}}|u_0, v_0, \alpha_\omega) \leqslant 0} f(\hat{\mathbf{z}}\,|u_0, v_0, \alpha_\omega)\, \mathrm{d}\hat{\mathbf{z}} \tag{5.8}$$

where u_0, v_0 are the mean wind speed and vehicle speed, respectively. α_ω is the wind angle, $P_f(u_0, v_0, \alpha_\omega)$ refers to the failure probability under the condition of u_0, v_0, α_ω, $f(\hat{\mathbf{z}}\,|u_0, v_0, \alpha_\omega)$ is the conditional probability density function over the failure domain $g(\hat{\mathbf{z}}|u_0, v_0, \alpha_\omega) \leqslant 0$.

Considering a railway vehicle running on curved track, the failure probability is influenced not only by the wind and vehicle speeds, but also by the track radius R and cant deficiency h_e, cf. Figure 4.28. Thus, the failure probability can be defined as

$$P_G = P_{f,curve} = P(u_0, v_0, \alpha_\omega, h_e, R) = \int\limits_{g(\hat{\mathbf{z}}|u_0, v_0, \alpha_\omega, h_e, R) \leqslant 0} f(\hat{\mathbf{z}}|u_0, v_0, \alpha_\omega, h_e, R)\mathrm{d}\hat{\mathbf{z}} \tag{5.9}$$

5.4 Methods to determine the failure probability

The failure probability of P_f or $P_{f,curve}$ can be computed either by analytical methods, e.g. first-order reliability method (FORM) [68] [164] or numerical methods like Monte-Carlo simulation with variance reduction (e.g. line sampling (LS) [123] and subset simulation (SS) [5]).

5.4.1 FORM method

FORM can be well utilized for reliability problems with linear-state functions and Gaussian-distributed random variables, cf. [68] [204]. It takes the minimal distance from the origin of the standard normal space to the limit state function $g(\hat{\mathbf{z}}) = 0$ as a reliability index [176]. An approximation of the failure probability P_f by FORM can be given by

$$P_f = \Phi\left(-\beta\right) \tag{5.10}$$

where

$$\beta = \sqrt{\left(\hat{\mathbf{z}}_{MPP}\right)^T \left(\hat{\mathbf{z}}_{MPP}\right)} \tag{5.11}$$

β refers to the reliability index and $\hat{\mathbf{z}}_{MPP}$ is the most probable point (MPP) which gives the highest probability among all the points in the failure domain $\mathbf{F}$.

It should be noted that FORM is an accurate procedure to determine the failure probability associated with linear or nearly linear limit state functions in the standard normal space. However, it appears to lack accuracy for nonlinear and high dimensional problems, cf. [176] [205]. In this case, other methods like Monte-Carlo and its variants have to be applied.

5.4.2 Monte-Carlo method

Monte-Carlo method determines the failure probability of an event by [176] [123]

$$P_f = P(\hat{\mathbf{z}} \in \mathbf{F}) = \int \mathrm{I}_{\mathbf{F}}(\hat{\mathbf{z}}) f(\hat{\mathbf{z}}) \mathrm{d}\hat{\mathbf{z}} \tag{5.12}$$

which is usually approximated as

$$P_f \approx \frac{1}{N} \sum_{i=1}^{N} \mathrm{I}_{\mathbf{F}}(\hat{\mathbf{z}}_i) \tag{5.13}$$

where N refers to the number of simulation times, $\hat{\mathbf{z}}_i$ are independent and identically distributed samples with a probability density function $f(\hat{\mathbf{z}})$, $\mathbf{F}$ refers to the failure domain and $\mathrm{I}_\mathbf{F}$ is an indicator function which can be defined as

$$\mathrm{I}_\mathbf{F} = \begin{cases} 1 & , \hat{\mathbf{z}} \in \mathbf{F} \\ 0 & , \hat{\mathbf{z}} \notin \mathbf{F} \end{cases} \tag{5.14}$$

Monte-Carlo method is robust and almost applicable to all kinds of reliability problems. The main disadvantage of it is its low efficiency to estimate small failure probabilities since it needs a large number of independent samples to achieve reasonable accuracy [6]. For this reason, variance reducing techniques have been developed, such as the importance sampling (IS) method [120] [150], the line sampling method and the subset sampling method [5].

5.4.3 Line sampling (LS) and Subset sampling

LS determines the failure probability by using the important direction $\mathbf{e}_\alpha$, which can be obtained by FORM. When the important direction is found, LS will be very efficient.

For each sample, LS decomposes the sampling direction into two directions [176] [202], i.e.

$$\hat{\mathbf{z}}_i = \left\|\hat{\mathbf{z}}_{i\|}\right\| \mathbf{e}_\alpha + \hat{\mathbf{z}}_{i\perp} \tag{5.15}$$

where $\hat{\mathbf{z}}_{i\|}$ is the component parallel to $\mathbf{e}_\alpha$ and $\hat{\mathbf{z}}_{i\perp}$ is the component perpendicular to $\mathbf{e}_\alpha$.

The failure probability P_f can be determined by increasing or decreasing $\left\|\hat{\mathbf{z}}_{i\|}\right\|$ of the initial sampling point until it reaches the limit state function. For N samples

$$P_f = \frac{1}{N}\sum_{i=1}^{N}\Phi(-\beta_i) \tag{5.16}$$

where β_i denotes the final length of the parallel component of each sampling point after several iterations.

Figure 5.3 gives an example of the sampling points by LS method in computing the failure probability of a vehicle running in strong crosswind. Since each sampling point has 8 coordinates, to show it clearly in a three-dimensional coordinate system, arbitrary three coordinates are chosen (here, take the gust amplitude A, gust duration T and aerodynamic coefficient C_{mx} for example). As can be seen, the sampling points are centered around the MPP, which shows the high efficiency of the LS method.

Besides LS, the SS method is also widely used in reliability analysis for small probability calculations, cf. [187] [179] and [5]. Its potential advantage is that it can avoid the simulation of rare events [6].

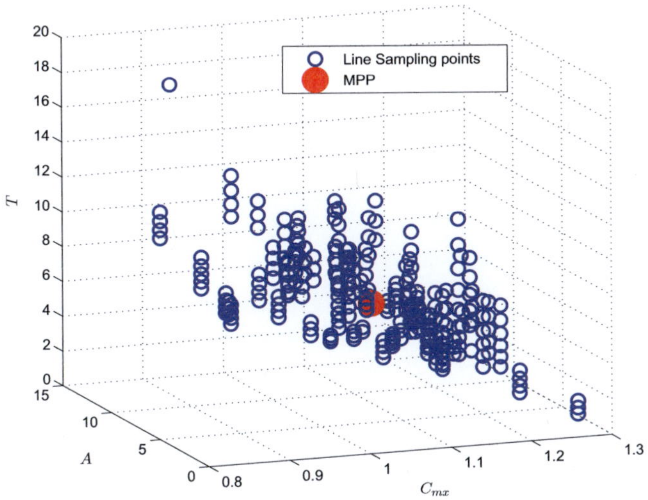

Figure 5.3: The distribution of line sampling points.

The main principle of SS is to calculate the probability in the target failure domain **F** by introducing a sequence of nested failure domains $\mathbf{F}_1 \supset \mathbf{F}_2 \supset \cdots \supset \mathbf{F}_n = \mathbf{F}$. Therefore,

$$\begin{aligned} P_f(\mathbf{F}) &= P(\mathbf{F}_n|\mathbf{F}_{n-1})P(\mathbf{F}_{n-1}) \\ &= P(\mathbf{F}_n|\mathbf{F}_{n-1})P(\mathbf{F}_{n-1}|\mathbf{F}_{n-2})P(\mathbf{F}_{n-2}) = \cdots = P(\mathbf{F}_1)\prod_{i=2}^{n} P(\mathbf{F}_i|\mathbf{F}_{i-1}) \end{aligned} \tag{5.17}$$

where $P_f(\mathbf{F})$ refers to the failure probability in the failure domain **F** and $P(\mathbf{F}_i|\mathbf{F}_{i-1})$ is the conditional failure probability in each nested failure domain.

5.5 Failure probability of the vehicle for a time interval

It is interesting to study the failure probability of a vehicle running with varying vehicle speed under nonstationary wind excitations in a relatively long journey during a time interval $t = [0 \;\; T_f]$. To calculate the failure probability of such a process, a link between the reliability analysis for the vehicle under a complete turbulent wind process and the mean gust shape scenario has to be identified. The complete turbulent wind process leads to a time-dependent reliability analysis with a large number of random variables while the mean gust shape scenario with only few random variables can be treated by means of static reliability procedures. These two solutions can be linked if the assumption holds that the failure events are rare and they always come along with an extreme gust [161].

Then, it is possible to expect that the failure events follow a discrete Poisson distribution with a failure rate λ^+, where

$$\lambda^+ = N^+(u_0)P_G(u_0, v_0, \alpha_\omega) \tag{5.18}$$

$N^+(u_0)$ is the occurrence frequency of the gusts over the level u_0 and $P_G(u_0, v_0, \alpha_\omega)$ is the failure probability for one gust event.

It should be noted that Equation (5.18) only refers to a stationary case. It is based on the assumption that the wind and vehicle speeds are constant and the wind angle does not change during the whole journey.

For a nonstationary case, the wind speed and vehicle speed as well as the wind angle change all the time, therefore the failure rate λ^+ has to be redefined as:

$$\lambda^+ = N^+(u(t))P_G(u(t), v(t), \alpha_\omega(t)) \tag{5.19}$$

Due to the varying vehicle and wind speeds, an exact numerical solution for $P_{f,T_f}(u(t), v(t), \alpha_\omega(t))$ is not so easy to obtain. One feasible way is to divide the track into several short sections, in each short track, the wind speed and the vehicle speed are thought to be constant. Following this, the global failure probability for a vehicle in a nonstationary scenario can be determined as follows

$$\begin{aligned} P_{f,T_f}(u(t), v(t), \alpha_w(t)) &= 1 - \prod_{i=1}^{n}(1 - P_{f,\Delta t_i}(u_0, v_0, \alpha_\omega)) \\ &= 1 - \prod_{i=1}^{n} \exp\{-\lambda_i^+ \cdot \Delta t_i\} = 1 - \exp\left\{-\sum_{i=1}^{n} \lambda_i^+ \cdot \Delta t_i\right\} \end{aligned} \tag{5.20}$$

where $P_{f,\Delta t_i}$ refers to the failure probability for each small section of the track and Δt_i is the corresponding time that the vehicle needs to pass the section.

The probability $P_{f,T_f}(u(t), v(t), \alpha_w)$ can be computed by Monte-Carlo method in principle. However, for such a nonstationary scenario, especially when the time period T_f is long, the Monte-Carlo based method is indeed time-consuming and inefficient. To circumvent this problem, the maximum-value method (MAX) and peak-over-threshold (POT) method are applied. According to the generalized-extreme-value (GEV) theory, the distribution of maximum values can be defined by the following equation [57] [79]

$$G(x) = \exp\{-[1 + \varsigma(\frac{x-b}{a})]^{-1/\varsigma}\} \quad , \quad 1 + \varsigma(\frac{x-b}{a}) > 0 \tag{5.21}$$

where a, b, ζ refer to the scale parameter, location parameter and shape parameter, respectively. To fit these three parameters, a large number of maximal values will be needed, few values would give a bad estimate. However, the MAX method can only give N largest values for N times independent simulations, which is relatively less. To avoid this drawback, the POT method is usually applied, which gives the asymptotic distribution as Pareto distribution [82] [167]

$$G(x) = 1 - \left(1 + \frac{\zeta x}{\beta}\right)^{-1/\zeta} \quad , \quad \zeta \neq 0 \tag{5.22}$$

where β, ζ are the scale and shape parameters. They can be determined based on the maximum likelihood estimation (MLE) [64] [89].

It should be noted that the choice of an appropriate threshold for the probability analysis is very important when applying the POT method. On the one hand, too large thresholds will yield a large variance for the estimator. On the other hand, if the threshold is too small, the approximation of of the excess distribution by Pareto distribution is also not so good since the bias of the POT quantile estimator will be no longer reliable, see [82]. In practice, the threshold is usually determined according to the confidence intervals of the estimated values (cf. Figure 6.6 in Chapter 6) or based on experience.

6 Nonstationary crosswind stability of railway vehicles

This chapter studies the crosswind stability of railway vehicles running under nonstationary crosswind on both straight and curved tracks. To begin with, results for the risk analysis of a vehicle running on straight track with varying vehicle speed will be presented in the following section. Strong crosswinds with or without nonstationary turbulence are considered. Besides, risk analysis has also been carried out for a vehicle running in a relatively long time interval and in an artificial storm. Then, in the second section, studies will focus on the stability of vehicles running on curved track. Track geometry has a great influence on crosswind stability. Regarding this, the influence of various design and track parameters are analyzed. In order to find the most influential parameters, a sensitivity analysis has also been undertaken.

6.1 Crosswind stability of vehicles running on straight track

6.1.1 Risk analysis for a vehicle running with varying speed

As already mentioned in Chapter 3, the wind speed and angle as well as the vehicle speed influence the relative wind speed and thus the wind forces and failure probabilities. Figure 6.1 presents an example of the failure probability P_G of a vehicle running on straight track at a constant vehicle speed $v_0 = 160$ km/h with different wind speeds and wind angles. It can be seen that high wind speeds and small wind angles will give rise to large failure probabilities.

When a nonstationary vehicle state is taken into account, i.e. the vehicle speed is not constant anymore, situations will become more complicated. To study the influence of varying vehicle speeds on the crosswind stability of railway vehicles, a risk analysis for the vehicle running with acceleration (AC) or deceleration (DE) has been carried out. Main methods for the risk analysis are based on FORM and LS, respectively. In addition, it is assumed that the acceleration a_c is equal to 1 m/s^2 for acceleration and -1 m/s^2 for deceleration. Based on the proposed method, it is possible to quantify the influence of various different parameters on the crosswind stability of vehicles.

Figure 6.2 compares the failure probability of a vehicle running under the excitation of nonstationary crosswind without wind turbulence at an initial vehicle speed $v_0 = 160$ km/h

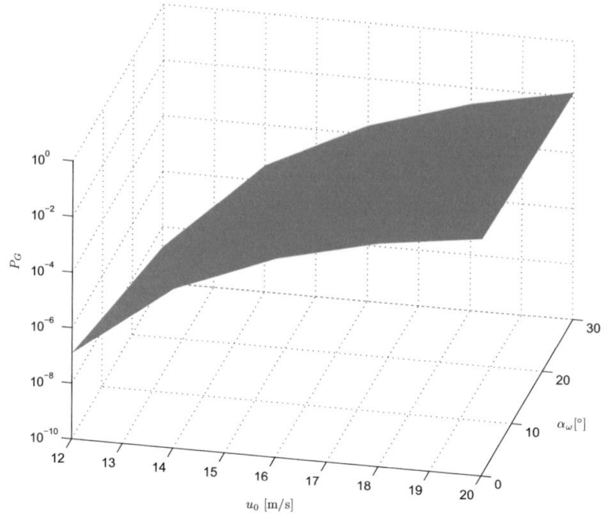

Figure 6.1: Failure probability of a vehicle running on straight track at $v_0 = 160$ km/h with different wind speeds and wind angles

with acceleration by FORM and LS, respectively. As can be seen, the two results match quite well. This good agreement makes it possible to compute the failure probability of vehicles running in strong crosswind by FORM instead of LS. However, it should be noted that FORM is only accurate for the case that the wind turbulence is not considered. When the wind fluctuations are taken into account, FORM will be not reliable any more due to the high dimensions, cf. Chapter 5.

Figure 6.3 displays the failure probability of a vehicle running in three different scenarios, i.e. the vehicle running at a constant speed, with acceleration and with deceleration, respectively. It is obvious that when the vehicle accelerates, the failure probability will be high. This is mainly caused by the higher relative wind speed. Although the relative wind angle $\beta_\omega(t)$ may also change due to the varying vehicle speed, its influence is not noticeable compared to the vehicle speed itself.

In Figure 6.4, the failure probability of a vehicle running with different initial vehicle speeds has been presented. Initial speeds for the vehicle are set as 160 km/h and 180 km/h, respectively. The results show again that the vehicle speed has an effect on crosswind stability. It can be imagined that on exposed sites where the vehicles may be blown over, speed limitations would be a very effective way to reduce the risk for wind-related accidents.

Moreover, to study the influence of wind turbulence on crosswind stability, a vehicle that runs at a constant vehicle speed or accelerates under the wind with or without turbulence has been investigated. As can be seen in Figure 6.5, the influence of wind turbulence on crosswind stability at low wind speeds is small. However, it becomes larger when the

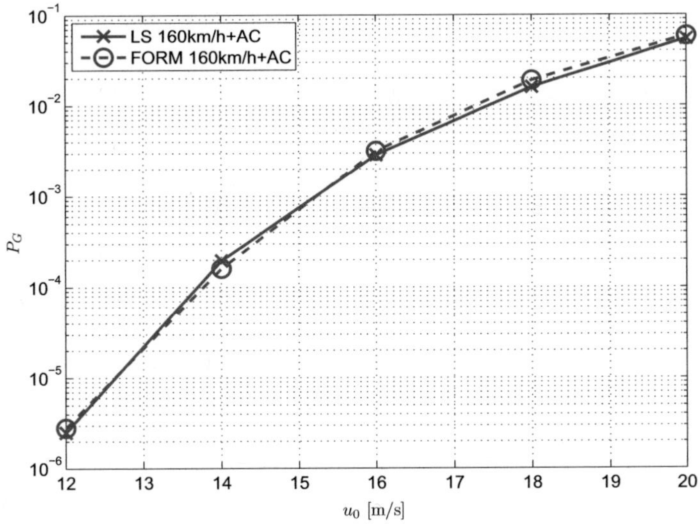

Figure 6.2: Failure probability of the vehicle at $v_0 = 160$ km/h with acceleration by FORM and LS methods

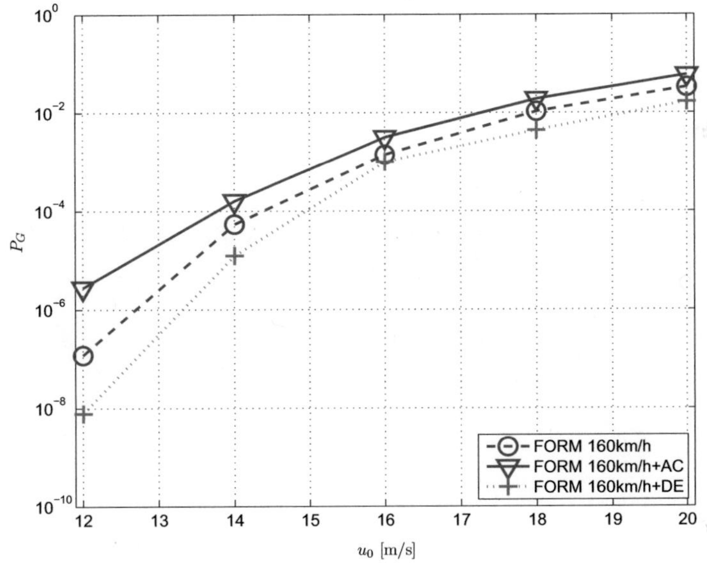

Figure 6.3: Failure probability P_G for initial vehicle speed $v_0 = 160$ km/h

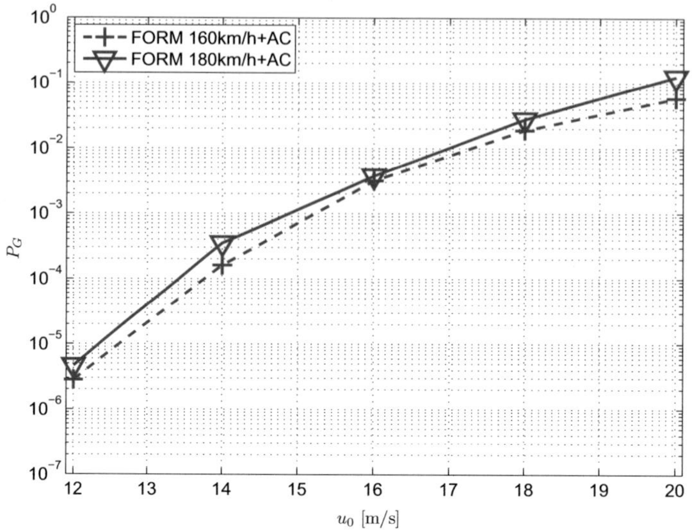

Figure 6.4: Failure probability P_G for vehicle speed $v_0 = 160, 180$ km/h with acceleration.

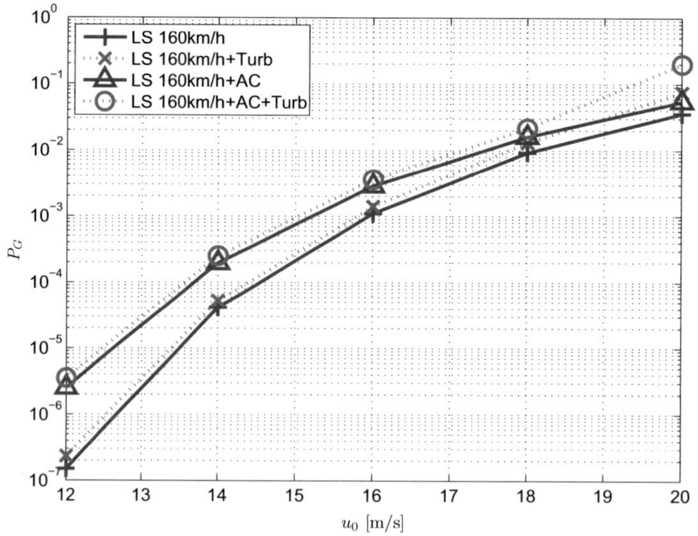

Figure 6.5: Failure probability P_G for vehicle speed $v_0 = 160$ km/h with or without wind turbulence.

mean wind speed is high. At the mean wind speed $u_0 = 20$ m/s, the influence is especially notable.

6.1.2 Failure probability of a vehicle for a time interval

In order to evaluate the failure probability of a vehicle running with varying vehicle speed in a relatively long journey, the vehicle running for a time interval under excitation of nonstationary crosswind has been investigated. It is assumed that the vehicle accelerates from an initial vehicle speed $v_0 = 160$ km/h to about 200 km/h during the time interval $T_f = 600$ s (10 min) and the wind turbulence is also taken into account.

Two methods: MAX and POT have been applied for the evaluation of failure probabilities, cf. Section 5.5. The MAX method takes all the maximum values (here, i.e. the maximum values of the wheel unloading change δQ, cf. Equation (5.1)) into account to fit the distribution parameters. However, one simulation can yield only one maximum value. For limited simulation times, the relevant random variables for MAX method may be insufficient. Therefore, the POT method, which takes all the values above a certain threshold into consideration, is usually applied.

Figure 6.6 shows an example of the estimated values of the shape parameter ζ for POT method with 95% confidence interval. From the figure, the confidence interval of ζ becomes larger when the threshold value $u_{\delta Q_{\max}}$ is higher. Since the threshold value can not be too large or too small (cf. Chapter 5), here, a value of $u_{\delta Q_{\max}}$ between 0.50 and 0.60 has been adopted.

Figure 6.7 and Figure 6.8 present the probability density function of $\delta Q_{\max}$ for the vehicle with initial velocity $v_0 = 160$ km/h by POT and MAX methods, respectively. The wind speed u_0 is set as 12 m/s and the threshold value for $\delta Q_{\max}$ is selected as $u_{\delta Q_{\max}} = 0.5$. As can be seen from the figure, the probability density of $\delta Q_{\max}$ exceeding the threshold value appears to decrease as an exponential mode. Both the Pareto fitting and Weibull fitting are consistent with the probability density of simulation results. The same good consistency also appears in Figure 6.8 by using GEV fitting and Gauss fitting with MAX method.

The probability density function of $\delta Q_{\max}$ under initial velocity $v_0 = 160$ km/h, $u_0 = 14$ m/s by POT and MAX methods are presented in Figure 6.9 and Figure 6.10, respectively. Compared to Figure 6.7 and Figure 6.8, it can be concluded that with the increase of vehicle speed and wind speed, the change of wheel unloading tends to be larger, which will give rise to a higher failure probability of the vehicle.

Failure probabilities of the vehicle running with or without acceleration during the time interval $T_f = 600$ s are shown in Figure 6.11. As can be seen, the probabilities obtained by POT and MAX methods show the same trend. The failure probability becomes high when the vehicle accelerates. Although the acceleration given here is small, the influence of it on failure probability is notable. However, it should be noted that the failure probability calculated here is a cumulative probability. Assume that the average occurrence rate of

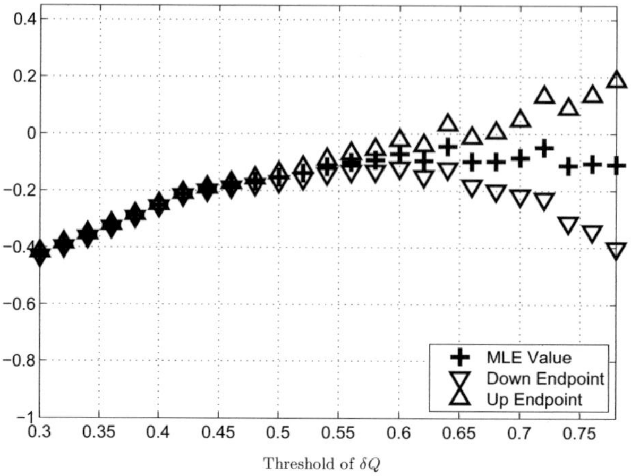

Figure 6.6: The estimated values of the shape parameter ζ for POT method with 95% confidence interval

the gust event is about 1/25s, then the occurrences of the gust event during $T_f = 600$ s are more than 20 times. Therefore, failure probabilities here are rather high. Another fact should be noticed that the POT method gives a relatively higher evaluation of the failure probability than the MAX method although the difference is not so large. This is probably because of the limited number of samplings. Compared to the MAX method, the POT method has a larger sampling space. For a limited sampling space, POT results may be more reliable.

6.1.3 Failure probability of a vehicle running in an artificial storm with acceleration

In this section, a train running in an artificial storm with acceleration is studied. The storm model is based on [98], which has also been used in the past for the analysis of crosswind stability of vehicles running with constant vehicle speed, cf. [74] [202]. Figure 6.12 shows a sketch of a train passing through the storm, where R_0 refers to the initial radius of the wind speed from the storm center, R_t is the radius at time t and v_c refers to the translation speed of the storm. d_0 and Δd_0 are the initial distance between the storm center and the vehicle in the horizontal and lateral directions, respectively.

For simplicity, it is assumed that the translational speed of the storm is constant during the simulation and the influence of the terrain to the storm is negligible. In addition, the train is considered to run on a straight track. Then, the gradient wind speed $V_g(R)$ of the

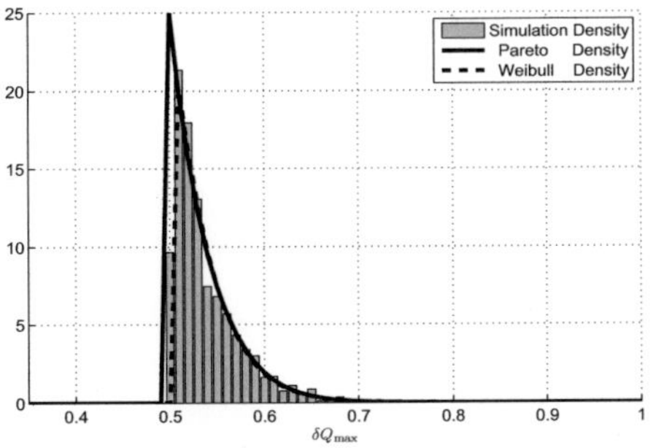

Figure 6.7: Probability density evaluation by POT method for $v_0 = 160$ km/h, $u_0 = 12$ m/s

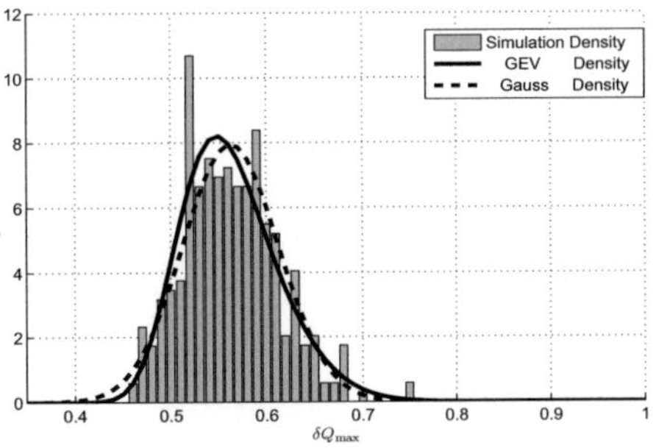

Figure 6.8: Probability density evaluation by MAX method for $v_0 = 160$ km/h, $u_0 = 12$ m/s

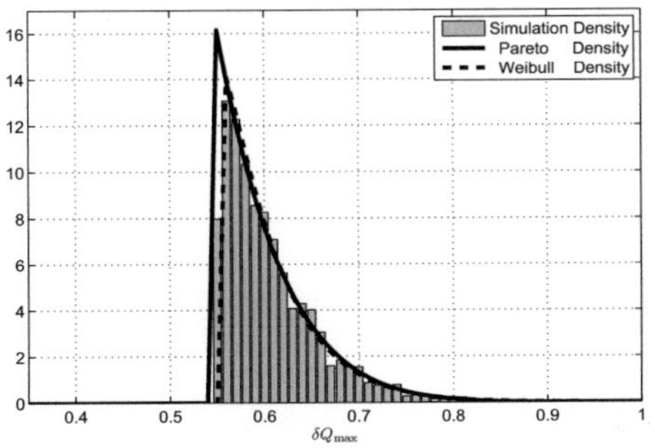

Figure 6.9: Probability density evaluation by POT method for $v_0 = 160$ km/h, $u_0 = 14$ m/s

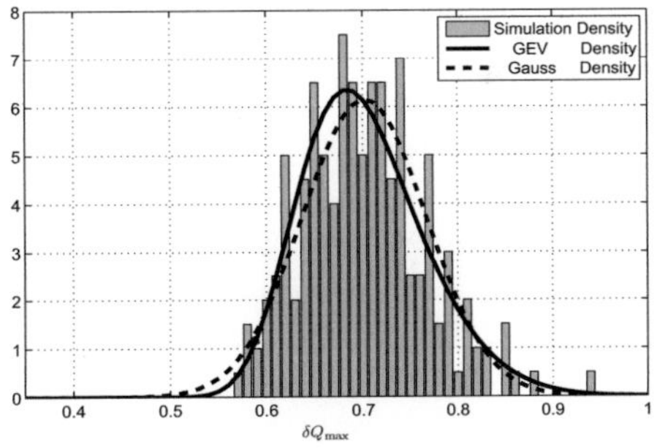

Figure 6.10: Probability density evaluation by MAX method for $v_0 = 160$ km/h, $u_0 = 14$ m/s

storm in a radial profile can be expresses as follows [74]

$$V_g(R) = \sqrt{\frac{\beta_s}{\rho_L} R_\beta (P_E - P_C) e^{-R_\beta} + \frac{{R_t}^2 {f_p}^2}{4}} - \frac{1}{2} R_t f_p \tag{6.1}$$

where

$$R_\beta = (\frac{R_{\max}}{R_t})^{\beta_s} \tag{6.2}$$

$$R_t = \sqrt{\left(d_0 - (v_0 t + \frac{1}{2} a_c t^2) \sin \alpha_2\right)^2 + \left(\Delta d_0 + (v_0 t + \frac{1}{2} a_c t^2) \cos \alpha_2 - v_c t\right)^2} \tag{6.3}$$

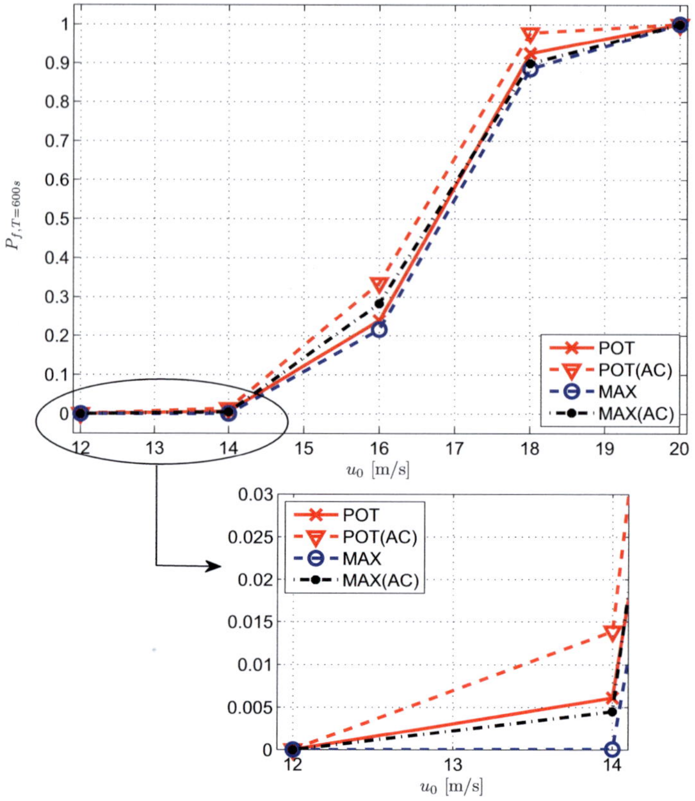

Figure 6.11: Time-dependent failure probability $P_{f,T=600\mathrm{s}}$ under different situations by extreme value method

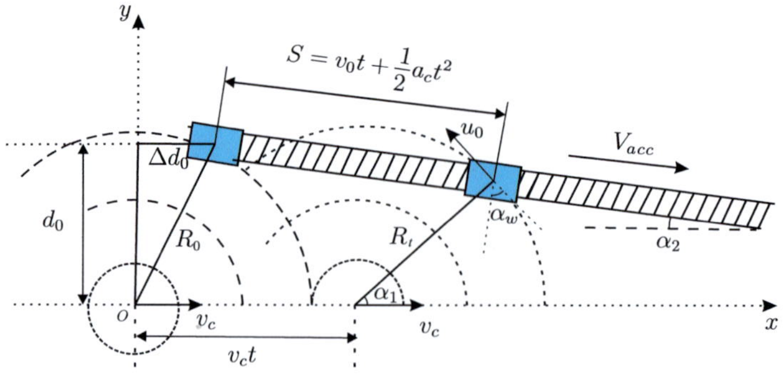

Figure 6.12: Sketch of the train passing through a storm

β_s is the radial shape parameter, ρ_L the air density, P_E the asymptotic environmental pressure, P_C the center pressure and $R_{\max}$ the radius of the maximum wind speed, f_p is the Coriolis parameter which can be determined by

$$f_p = 2\omega_e \sin(\alpha_e) \tag{6.4}$$

in which ω_e is the angular speed of the earth and α_e is the latitude.

According to [74], the gradient wind speed $V_g(R)$ can be converted to the wind speed at 10m height above the ground by a constant coefficient

$$u_{10} = 0.865 V_g(R) \tag{6.5}$$

By the logarithmic law [203] [119], the wind speed at a reference height z can be obtained by

$$u(z) = 0.19(\frac{z_0}{0.05})^{0.07} u_{10} \ln(\frac{z}{z_0}) \tag{6.6}$$

The corresponding wind angle α_ω acting on the vehicle can be expressed as follows

$$\alpha_\omega(t) = \alpha_1 + \alpha_2 = \arctan\left(\frac{d_0 - (v_0 t + \frac{1}{2}a_c t^2)\sin\alpha_2}{\Delta d_0 + (v_0 t + \frac{1}{2}a_c t^2)\cos\alpha_2 - v_c t}\right) + \alpha_2 \tag{6.7}$$

Assume that the vehicle runs at an initial vehicle speed $v_0 = 160$ km/h with acceleration $a_c = 0.02$ m/s^2. Main parameters related to the storm are defined as in [119] [202]: $R_{max} =$ 20 km, $d_0 = 15$ km, $\Delta d_0 = 5 \times 10^4$ m, $\alpha_e = 45°$, $P_C = 97000$ N/m^2, $P_E = 101500$ N/m^2, $\alpha_2 = 5°$, $v_c = 15$ m/s. The failure probability of the vehicle as a function of time t can be obtained.

Figure 6.13 represents the mean wind speed that acts on the vehicle with or without acceleration as a function of time t. As can be seen, when the vehicle accelerates, the wind speed acting on the vehicle becomes smaller. This is mainly because of the change of the distance between the vehicle and the storm center. In the given scenario, when the vehicle accelerates, the relative distance between the vehicle and the storm center will become longer. As can be imagined, longer distance can lead to a smaller wind speed, this can also be deduced from Equations (6.1), (6.5) and (6.6). In Figure 6.14, corresponding wind angles acting on the vehicle are displayed.

The failure probability of the vehicle running with or without acceleration in the above storm scenario has been shown in Figure 6.15. As can be seen, the vehicle with acceleration has a larger failure probability than that without acceleration, although the wind speed acting on the vehicle with acceleration is smaller. Compared to the decreased wind speed caused by the acceleration, the increased vehicle speed leads to a higher relative wind speed. This implies again that the vehicle speed has a great influence on the crosswind stability of vehicles.

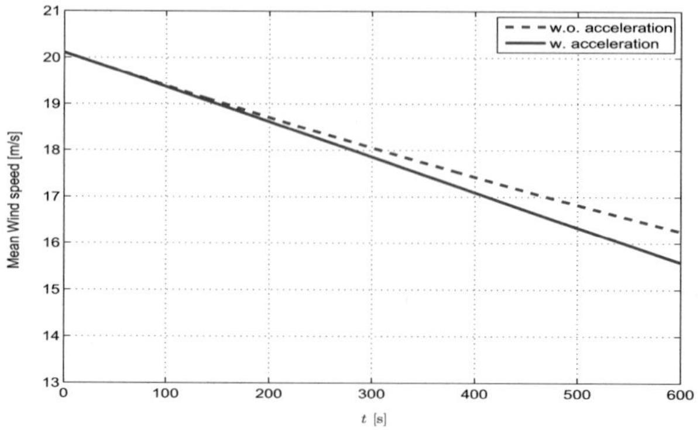

Figure 6.13: Mean wind speed acting on the vehicle in a storm scenario

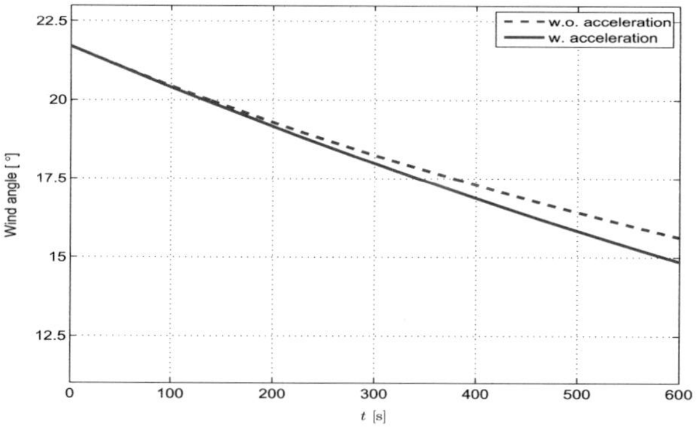

Figure 6.14: Corresponding angle of the mean wind speed

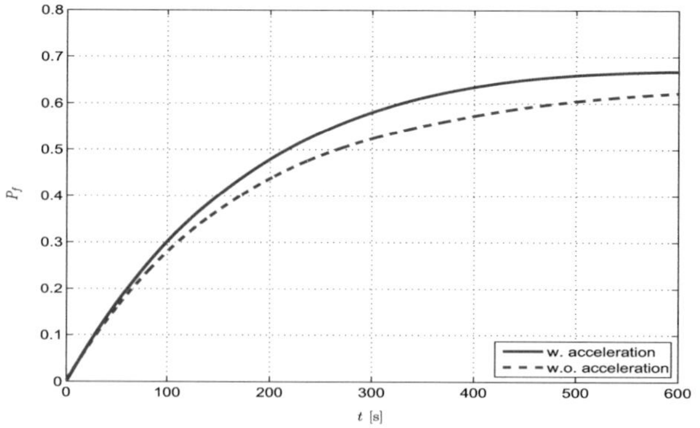

Figure 6.15: Failure probability of the vehicle running with or without acceleration in a storm scenario

6.2 Crosswind stability of vehicles running on curved track

6.2.1 Risk analysis

In this section, risk analysis for the vehicle running on curved track under nonstationary crosswind excitation has been carried out. The influence of different curve radii and cant deficiencies on the crosswind stability is studied and compared.

Figure 6.16 shows an example of the curved track line in a two-dimensional coordinate system. It contains three parts: the straight part, the transition part and the circular part. A simulation model of the vehicle running on curved track in Adams can be seen in Figure 6.17.

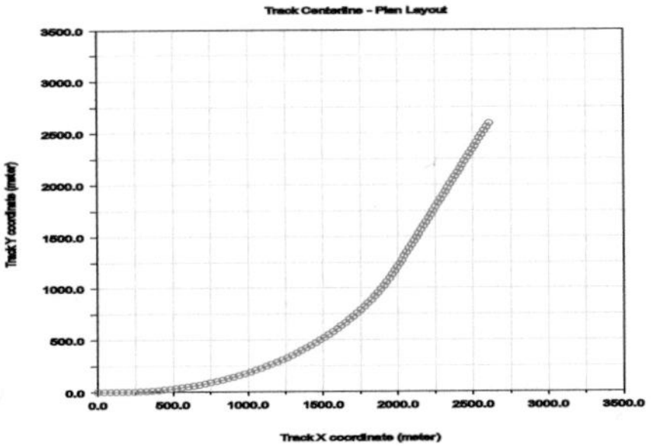

Figure 6.16: An example of the real curved track in a global axis.

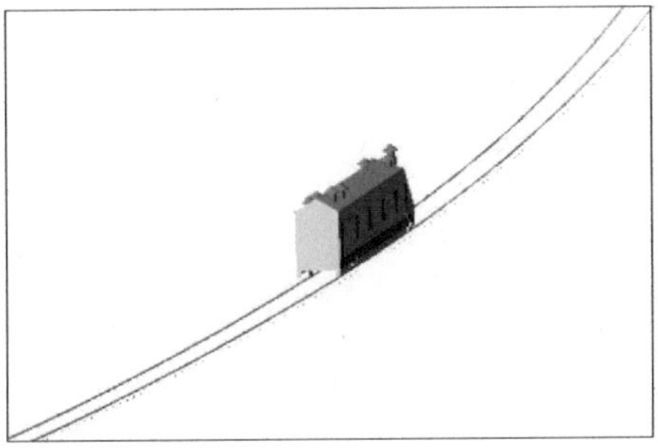

Figure 6.17: Simulation model of the vehicle running on the curved track in Adams.

Figure 6.18 shows the failure probability of a vehicle running at different velocities on curved track with $R = 2000$ m and $h_t = 100$ mm by FORM. It is obvious that the failure probability becomes high when the mean vehicle speed increases. For curved tracks, the vehicle speed also plays an important role on the crosswind stability.

Without considering the wind turbulence, a comparison of the failure probability for a vehicle running on curved track based on FORM and LS is represented in Figure 6.19. The wind is assumed to blow from the inner side of the track (cf. Figure 4.28 (*b*)). As can be seen, the results obtained from the two methods match quite well. This good agreement makes it also possible to use FORM instead of LS for the risk analysis of vehicle running on curved track under strong crosswinds without nonstationary turbulence.

When the nonstationary wind turbulence is taken into consideration, the failure risk of the vehicle will become larger. This can be seen in Figure 6.20. As is shown, the nonstationary wind turbulence has an influence on the crosswind stability. It should be noted again that when the turbulent fluctuations are considered, the FORM analysis will not be possible any more and LS has to be implemented instead.

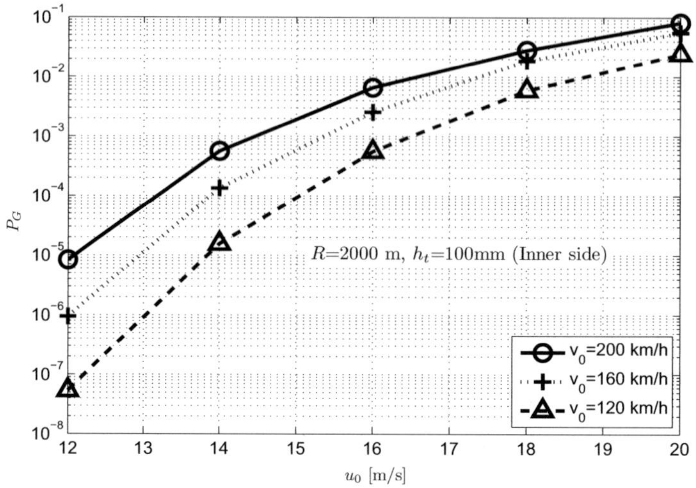

Figure 6.18: Failure probability P_G for different vehicle velocities

Figure 6.21 represents the influence of wind directions (from outer side or inner side) on the failure probability of a vehicle running on curved track with $R = 2000$ m, $h_t = 100$ mm. It can be seen that the wind direction has a great influence on the crosswind stability of vehicles. Winds coming from the inner side, which have the same direction with the unbalanced acceleration, will increase the probability of failure.

Figure 6.22 shows the failure probability calculated by FORM of a vehicle running on curved track with different horizontal radii. The wind fluctuations are not taken into account and the track cant h_t is set as 100 mm. In order to display the beneficial effect of the lateral acceleration, wind direction from the outer side (i.e. the centrifugal direction, cf. Figure 4.28 (a)) is studied here. It can be seen that with the increase of the curve radius, the lateral acceleration and the failure probability P_G decrease. For a large curve radius, when the lateral acceleration in the track plane a_y is nearly 0, the failure probability of the vehicle caused by the wind is almost the same as that of the vehicle running on straight track, which can be seen in Figure 6.23.

Figure 6.24 compares the failure probability of the vehicle running on curved track with different track cants under the wind excitation without turbulence. The radius of the curved track is $R = 2000$ m. Wind excitations are from the inner side. As is shown in the figure, the track cant has influence on the vehicle stability. For a fixed track radius, track cant changes the lateral acceleration of the vehicle. According to Equation (4.40), the corresponding lateral accelerations for track cant $h_t = 100$ mm and $h_t = 160$ mm are $a_y = 0.3321$ m/s^2 and $a_y = -0.0633$ m/s^2, respectively. On the one hand, when $a_y > 0$, the wind coming from the outer side will increase the unbalanced force and thus makes the failure risk of the vehicle larger. On the other hand, when $a_y < 0$, the crosswind can offset part of the unbalanced force and the failure probability will become smaller.

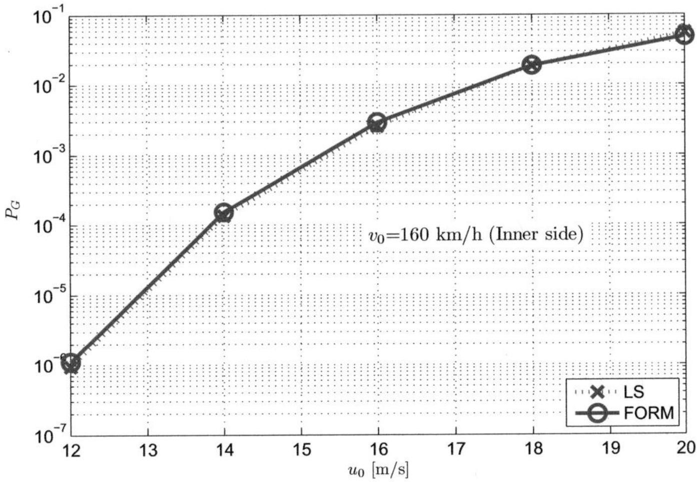

Figure 6.19: Comparison of the failure probability based on LS and FORM methods with $R = 2000$ m, $h_t = 100$ mm

In Figure 6.25, the influence of track irregularities on the failure probability is represented. When the track irregularity is taken into account, a sliding mean has to be applied for the wheel forces so that the critical wheel unloading can be ensured to last over a sufficiently long time. It can be seen that track irregularities have a visible, but not so large influence on the failure probability.

Next, Table 6.1 gives the detailed values of vehicle failure probability P_f with 98% confidence interval under the condition: $R = 2000$ m, $h_t = 100$ mm, $v_0 = 160$ km/h. The wind comes from the inner side. As can be seen, the width of the confidence interval of P_f is small, which shows a good precision of the estimation. Besides, failure probabilities for the vehicle running on curved track with irregularities and nonstationary wind turbulence under the same condition are also listed. It is shown that the influence of nonstationary wind turbulence on the failure probability is larger than that of track irregularities.

Table 6.1: Failure probability of the vehicle under the condition: $R = 2000$ m, $h_t = 100$ mm, $v_0 = 160$ km/h

u_0 [m/s]	12	14	16	18	20
P_f	9.41×10^{-7}	1.34×10^{-4}	2.55×10^{-3}	1.86×10^{-2}	5.53×10^{-2}
$P_{f,98\%}^{upper}$	9.55×10^{-7}	1.34×10^{-4}	2.60×10^{-3}	1.89×10^{-2}	5.65×10^{-2}
$P_{f,98\%}^{lower}$	9.26×10^{-7}	1.34×10^{-4}	2.50×10^{-3}	1.85×10^{-2}	5.42×10^{-2}
$P_{f,irr}$	1.47×10^{-6}	1.80×10^{-4}	3.15×10^{-3}	1.89×10^{-2}	5.61×10^{-2}
$P_{f,turb}$	2.42×10^{-6}	3.20×10^{-4}	5.30×10^{-3}	3.33×10^{-2}	1.36×10^{-1}

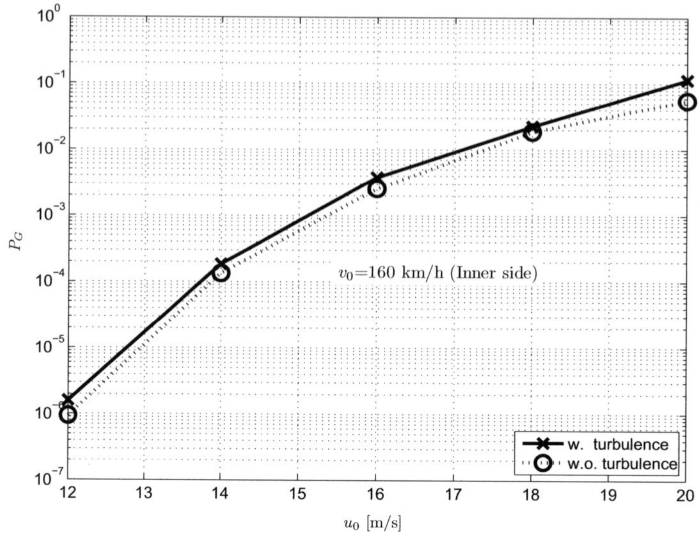

Figure 6.20: Influence of nonstationary wind turbulence on the failure probability of the vehicle running on curved track, $R = 2000$ m, $h_t = 100$ mm

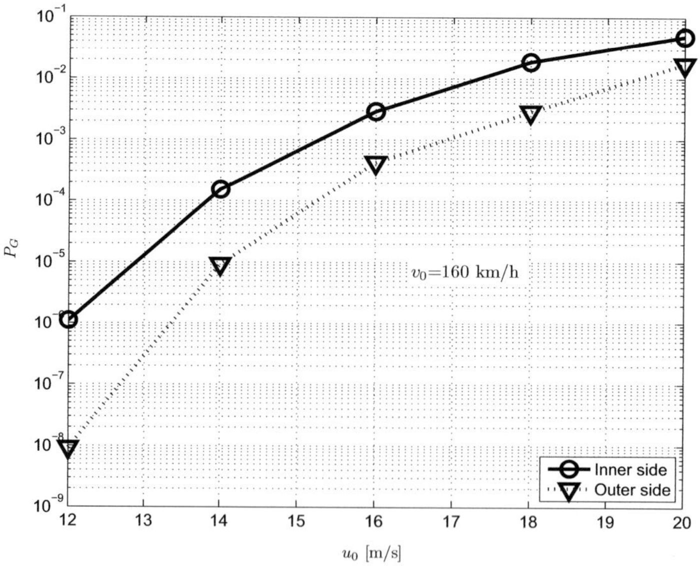

Figure 6.21: Influence of the wind direction on the failure probability of the vehicle running on curved track, $R = 2000$ m, $h_t = 100$ mm

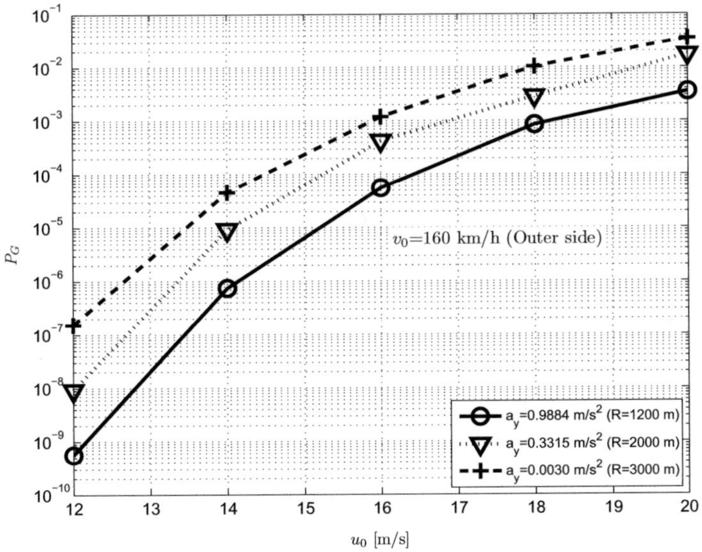

Figure 6.22: Failure probability of the vehicle running on curved track with different curve radii, $h_t = 100$ mm

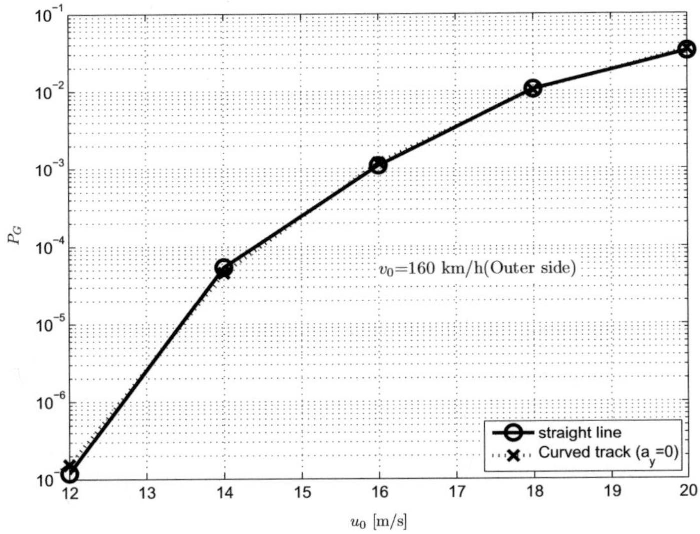

Figure 6.23: Comparison of the failure probability on straight track and curved track ($R =$ 3000 m)

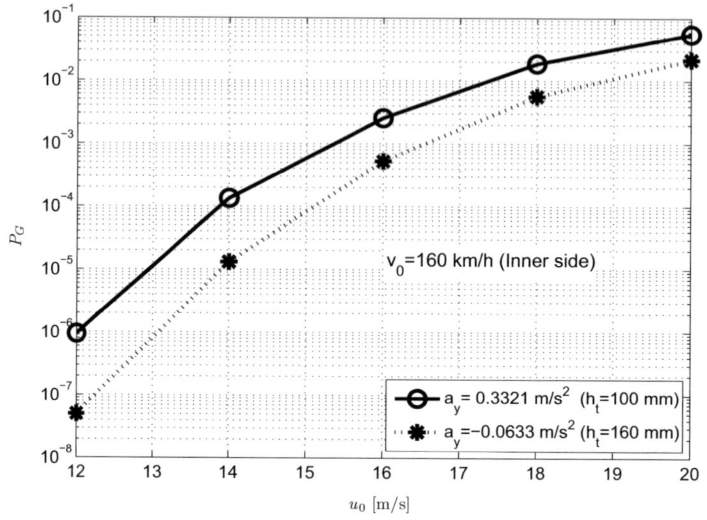

Figure 6.24: Influence of the cant on the failure probability of the vehicle under the wind excitation without turbulence

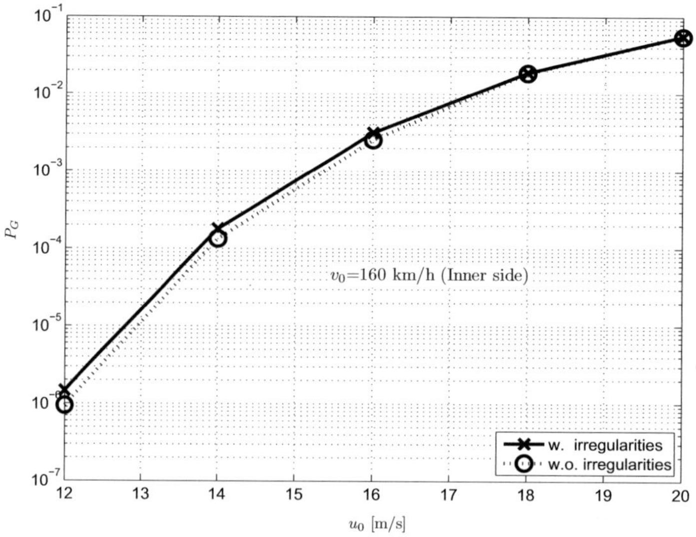

Figure 6.25: Failure probability of the vehicle running on the track with or without irregularities, $R = 2000$ m, $h_t = 100$ mm

6.2.2 Sensitivity analysis

In order to study the influence of input variables on the failure probabilities of the vehicle, the Morris method [173] [146] and the correlation-analysis method [131] are used. The Morris method evaluates the importance and the nonlinear influence of the random variables qualitatively. It starts from a random point and then computes the finite difference by a random walk by varying always only one of the n variables at a time. It can be defined as follows:

$$e_i = \frac{g(\hat{z}_1, \hat{z}_2, \cdots, \hat{z}_i + \Delta\hat{z}_i, \cdots \hat{z}_n) - g(\hat{\mathbf{z}})}{\Delta\hat{z}_i} \tag{6.8}$$

where $g(\hat{\mathbf{z}})$ and $\hat{z}_i$ are the limit state function and the stochastic variables, respectively. For n random walks, the mean value μ_i and the variance σ_i^2 are

$$\mu_i = \frac{1}{n}\sum_{k=1}^{n} e_{i,k}\,, \quad \sigma_i^2 = \frac{1}{n-1}\sum_{k=1}^{n}(e_{i,k}-\mu_i)^2 \tag{6.9}$$

Likewise, correlation-analysis can also help to get an insight into the sensitivity of variables of a stochastic system. Consider the vector $Y = [\hat{\mathbf{z}},\, g(\hat{\mathbf{z}})]^T$, the correlation coefficients can be defined as

$$\gamma_{i,j} = \frac{E\left[(Y_i - \mu_{Yi})(Y_j - \mu_{Yj})\right]}{\sqrt{[E(Y_i - \mu_{Yi})^2 E(Y_j - \mu_{Yj})^2]}} \tag{6.10}$$

where $Y_i \in Y = [\hat{\mathbf{z}}, g(\hat{\mathbf{z}})]^T$. As the variables $\hat{\mathbf{z}}$ are uncorrelated standard normal variables, only the last column of Y, i.e. the limit state function $g(\hat{\mathbf{z}})$ is of interest. The correlation coefficients $\gamma_{i,j} \in [-1, 1]$ describe the level of correlation, for which, -1 or 1 means total correlation and 0 means no correlation at all.

In Figure 6.26, a sensitivity analysis for the vehicle running on curved track with $v_0 =$ 160 km/h, $u_0 = 16$ m/s (direction: outer side) based on Morris method has been represented. The mean value μ indicates the importance of the variable to the system while the variance σ is a measure of the nonlinear effect. It can be seen that the gust amplitude A is the most influential parameter that has great impact on the crosswind stability of vehicles. Besides, the aerodynamic coefficient C_{mx} and the gust duration T also play an important role in the failure probability, although the influences of them are not so large compared to that of A.

A same sensitivity analysis for the vehicle-wind system under the initial velocity $v_0 =$ 160 km/h and $u_0 = 16$ m/s (direction: inner side) is shown in Figure 6.27, in which the

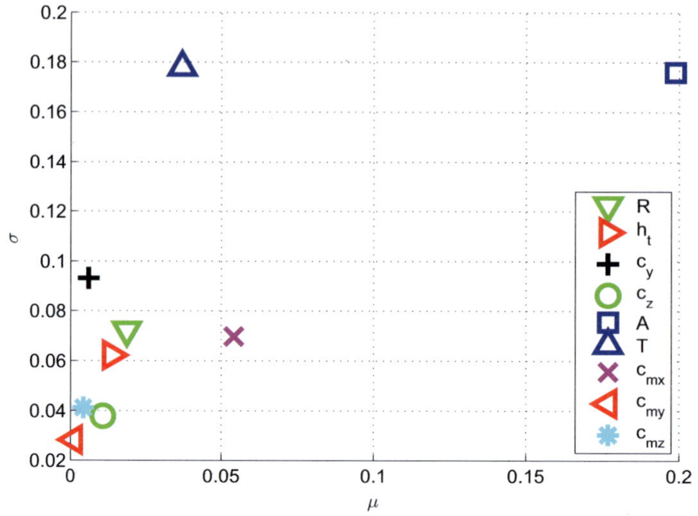

Figure 6.26: Sensitivity analysis based on Morris method for v_0 = 160 km/h and u_0 = 16 m/s. Wind direction: outer side.

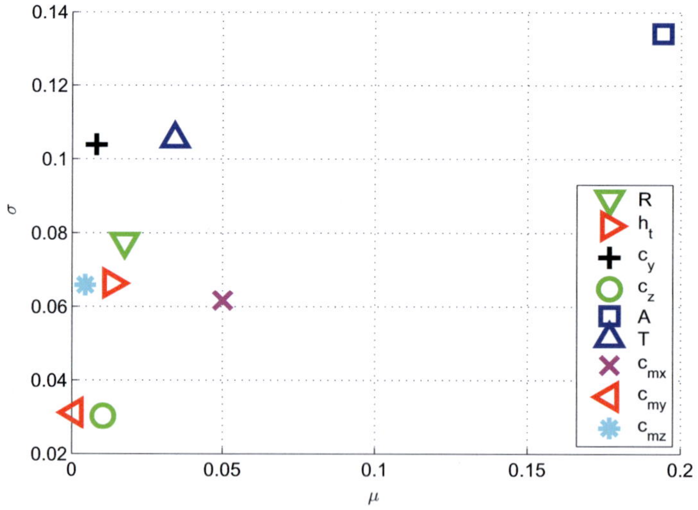

Figure 6.27: Sensitivity analysis based on Morris method for v_0 = 160 km/h and u_0 = 16 m/s. Wind direction: inner side.

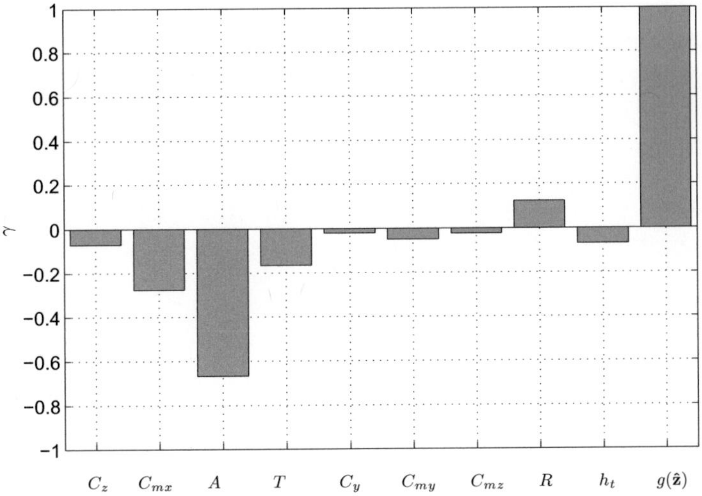

Figure 6.28: Sensitivity analysis based on correlation analysis for $v_0 = 160$ km/h and $u_0 = 16$ m/s Wind direction: outer side.

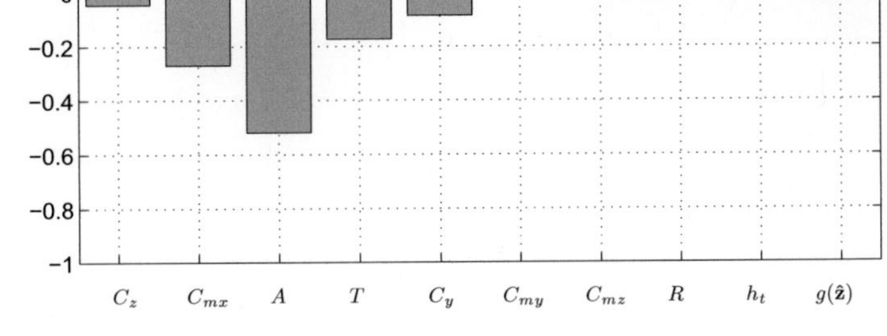

Figure 6.29: Sensitivity analysis based on correlation analysis for $v_0 = 160$ km/h and $u_0 = 16$ m/s. Wind direction: inner side.

most influential parameters are still the gust amplitude A, the aerodynamic coefficient C_{mx} and the gust duration T.

Figure 6.28 and Figure 6.29 present the results of sensitivity analysis for the above two scenarios based on correlation-analysis method. The correlation coefficients of each variable with respect to the given limit state function $g(\hat{\mathbf{z}})$ are displayed.

As expected, both figures show that the variables of A, C_{mx} and T have a relatively larger correlation coefficient with respect to $g(\hat{\mathbf{z}})$ compared to other variables, which indicates that these variables are more influential to the system. The result shows a good agreement with that obtained by Morris method.

7 Risk analysis for road vehicles under strong crosswind

Strong crosswinds have a great influence on the safety of road vehicles. They may cause the vehicle to overturn or undergo course deviations. Besides, at exposed locations (e.g. embankments and long bridges), the wind effect may be magnified by topographical features. Therefore, the safety analysis for road vehicles under strong crosswind is important.

So far, many methods (e.g. electronic stability aids (ESA), vehicle design improvement and driver training) have been proposed to improve the vehicle stability under strong crosswind [163], with the aim that the wind-related accidents can be reduced. However, these methods are usually costly and need time to be realized.

In practice, one way to ensure the safety of vehicles running under strong crosswind is to alert the driver to take certain actions (e.g. a signaling program with wind warning or vehicle speed limitation). In the case of extreme crosswinds, the traffic would be partly closed or even fully closed. In addition, it would be also possible to decrease the wind speed to keep the vehicle safe, for instance, a wind fence can be built up by the side of the road to reduce the impact of the wind. Although a lot of effort have been spent on preventing accidents of vehicles under strong crosswind [163], the study of crosswind stability is a great challenge. On the one hand, wind-reducing countermeasures (e.g. wind fence) are usually very expensive and cannot be widely implemented [143]. On the other hand, most of the earlier studies concerning crosswind stability of road vehicles are based on a deterministic wind model for a specific type of vehicle (e.g. a truck [90] [55], a bus [113], a car with trailer [138]). The determination of critical vehicle speeds for a signaling program based on the deterministic wind model is hard to reflect the real effect of the natural wind on road vehicles. Realistic analysis for crosswind stability has to take a stochastic crosswind model into consideration.

This chapter studies the crosswind stability for different road vehicles with the stochastic wind model. Gust amplitude A and gust duration T as well as the aerodynamic coefficients are considered as random variables. Besides, in order to be more realistic, actual measured wind data are used to calibrate the gust model. The vehicles are classified into five categories (i.e. car, truck, bus, van and car-trailer, cf. Chapter 4). For each vehicle type, the corresponding aerodynamic coefficients are identified.

Due to the existence of random variables, a risk analysis has to be carried out and failure probabilities for each vehicle type under various conditions have to be computed. Moreover, the influences of different parameters on the crosswind stability of vehicles are studied. The simulation allows to base a signaling program for traffic influence on a risk level.

7.1 Vehicle safety analysis and influence factors

As mentioned before, the safety analysis for road vehicles should consider not only the vehicle itself but also the influence that the vehicle may bring to the whole traffic, since the real traffic is usually very complicated. For example, when a vehicle runs on a two-lane road in strong crosswind, the driver has to control the vehicle to resist the wind so that the vehicle can be stable. At the same time, the driver needs to keep the vehicle away from the adjacent lane to avoid a crash. These conditions require a more strict constraint for the vehicle behavior. Therefore, several failure modes such as overturning, side slip and yaw angle have to be considered.

Totally speaking, there are many factors that influence the safety of road vehicles running in strong crosswinds, e.g. wind speed, wind direction and vehicle speed, etc.. It is intuitive that higher wind speeds with smaller wind angles (perpendicular to the running direction of the vehicle) may have a larger influence on the vehicles and tall vehicles would have a higher risk of overturning.

Moreover, the driver plays an important role in the crosswind stability of vehicles. Compared to an experienced driver, the inexperienced driver may not be able to control the vehicle well in bad weather conditions (e.g. rain, snow, ice, fog). Besides, the influence of road conditions on crosswind stability should be also noted. Wet road may increase the failure risk of road vehicles under strong crosswinds and sometimes may change accident modes of vehicles (e.g. from overturning to side slip).

Another thing should be mentioned that when a wind speed is extremely high, the vehicle may face a hazard no matter what speed it runs at. This means that in such a wind scenario, even if the vehicle speed is zero, the vehicle will still have a risk of failure. For example, a light-high truck may be blown over by an extremely strong crosswind even though it is static. Such cases seldom happen in reality and the corresponding failure probability of vehicles can be considered as 1.

7.2 Realistic wind model based on measured wind data

In order to obtain a more realistic assessment for the crosswind stability of vehicles, wind data measured on exposed sites with a constant time step in one year are used to calibrate the gust model. The PDF of gust amplitudes are analyzed based on the data.

Figure 7.1 and Figure 7.2 show an example of the PDF for gust amplitudes based on wind data measured at two different places, respectively. Comparing the two figures, a higher value of the PDF at larger gust amplitudes can be found in Figure 7.2, which implies that the wind speed at place 2 is generally larger than that at place 1. In addition, from the two figures, it can be seen that the Weibull PDF has a good fitting to the PDF of actual gust amplitudes.

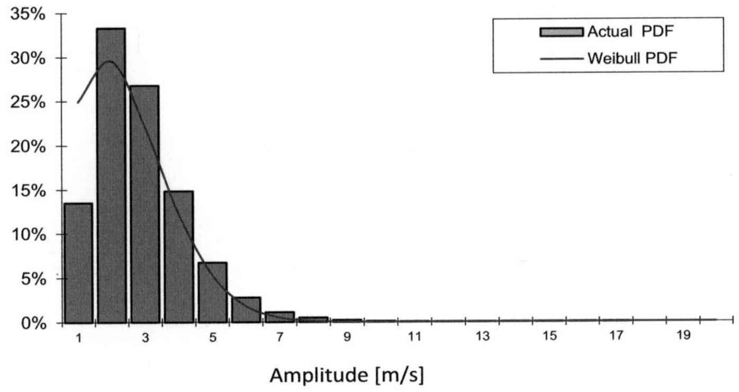

Figure 7.1: The PDF of gust amplitudes based on wind data measured at place 1

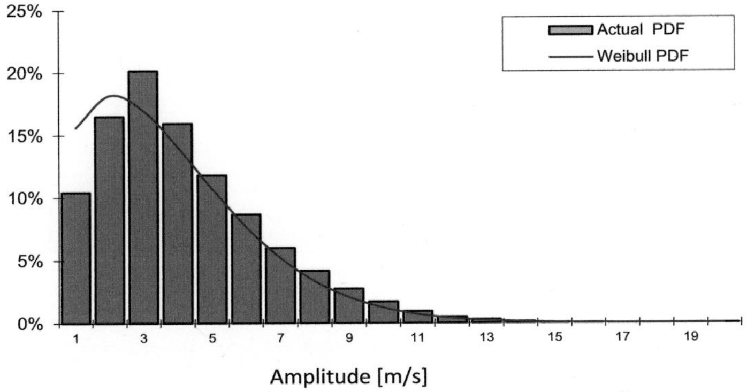

Figure 7.2: The PDF of gust amplitudes based on wind data measured at place 2

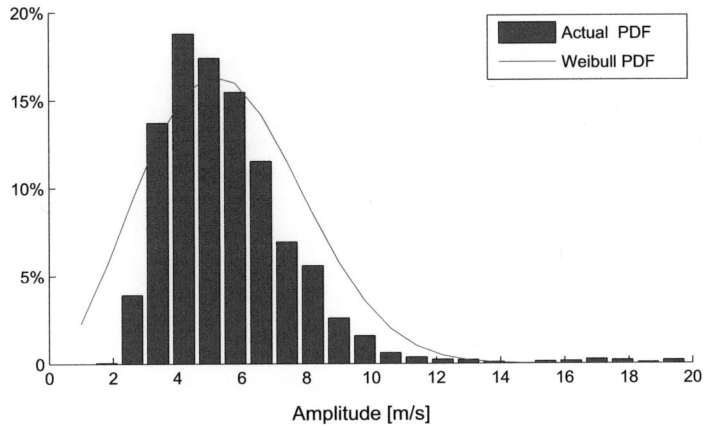

Figure 7.3: The PDF of gust amplitudes at mean wind speed $u_0 = 19$ m/s

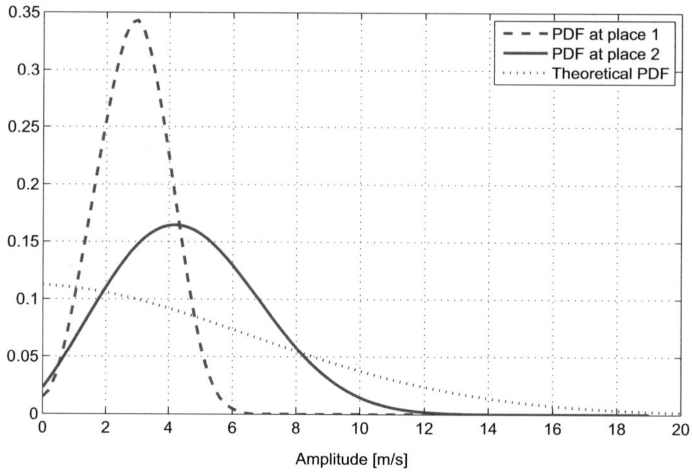

Figure 7.4: Comparison for the PDF of the gust amplitude at mean wind speed $u_0 = 19$ m/s

Figure 7.3 gives an example of the PDF for gust amplitudes at mean wind speed $u_0 = 19$ m/s at place 2. As can be seen, the gust amplitudes at $u_0 = 19$ m/s are mainly concentrated between 3 m/s and 9 m/s. A comparison of the PDF for gust amplitudes based on measured wind data and theoretical analysis (cf. Equation (2.74)) at $u_0 = 19$ m/s can be seen in Figure 7.4. Compared to the PDF of gust amplitudes at place 1, the PDF of gust amplitudes at place 2 is relatively wide, which again indicates that the wind speed at place 2 is in general higher than that at place 1. To be more conservative, in this work, unless specified, the wind speeds measured at place 2 are applied to calibrate the theoretical gust model.

7.3 Failure evaluation and risk analysis

In order to carry out the risk analysis, failure criteria as well as performance functions have to be defined first, cf. Chapter 5. For each vehicle class, a worst case vehicle model, i.e. an empty vehicle has been taken into account. The corresponding aerodynamic coefficients for each type of vehicle are given in the appendix.

The risk analysis may be conservative in several aspects. Firstly, it is assumed that the driver controls the vehicle only with the steering wheel, there is no braking or other assistant control program (e.g. the electronic stability program (ESP), active anti-roll suspension, etc.). Secondly, only the empty vehicle is considered here and the wind direction is assumed to be the same during a simulation (i.e. α_ω is constant), which in reality is not always the case, the vehicle is usually with load and the wind angle always changes due to the nonstationary character of the wind as well as due to the varying vehicle state

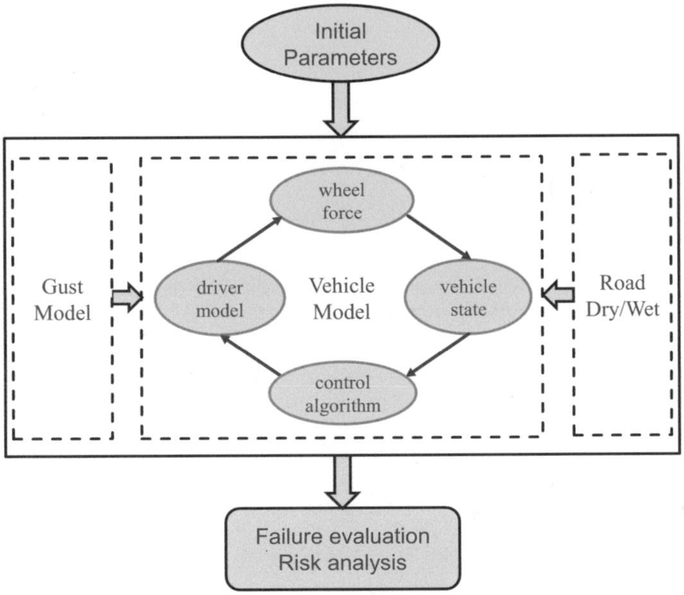

Figure 7.5: A sketch of the simulation program for road vehicles running in the crosswind

especially when the steering angle is operated. Thirdly, a driver will concentrate more on the driving in a gusty wind scenario.

Figure 7.5 shows a sketch for the simulation program. As can be seen, the simulation system mainly consists of three parts: the gust model, the vehicle model and the road model. In terms of the vehicle model itself, there are several subsystems, including the driver, the wheel and the vehicle body (vehicle state). They are fully coupled and should not be considered separately.

Figure 7.6 depicts the failure probability of a truck running on dry road at $v_0 = 60$ km/h with different wind excitations. Not surprisingly, the gust model calibrated with the wind data measured at place 2 causes a much higher failure probability due to the higher relative wind speed. It should be mentioned that, here and in the following, unless otherwise stated, the default wind angles are set as $\alpha = 0°$, i.e. the wind direction is perpendicular to the vehicle.

In Figure 7.7, failure probabilities of the truck running at different vehicle speeds on dry road are represented. The failure probability of the truck becomes larger with the increase of vehicle speed. The relative wind angle may also change due to the increase of vehicle speed. However, it has no noticeable influence on the failure probability compared to the vehicle speed itself. Similar results can be also found in Figure 7.8 for a passenger car. Compared to the passenger car, the truck has a larger failure probability at the same

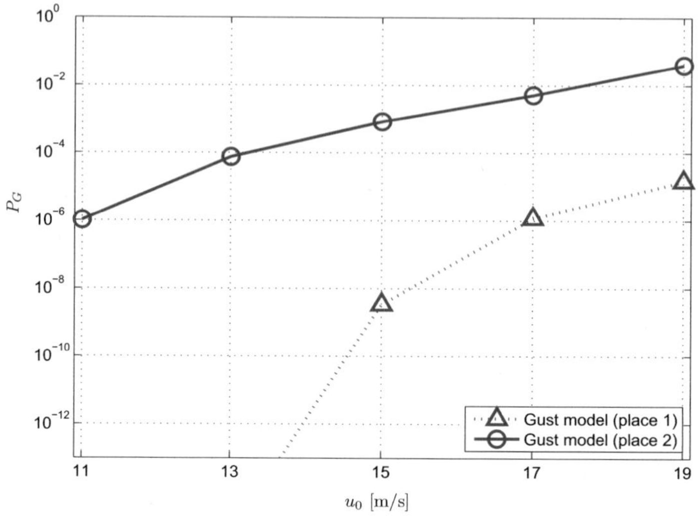

Figure 7.6: Failure probability of a truck running at v_0 = 60 km/h on dry road with different wind models

vehicle speed, which indicates that the truck is more sensitive to the crosswind than the passenger car.

Figure 7.9 displays the failure probability of different vehicles running on dry road at $v_0 =$ 60 km/h. The failure probability of the truck running under strong crosswind excitation with $\alpha_\omega = 30°$ has been also displayed because it is found that the truck will have a much higher failure probability at $\alpha_\omega = 30°$ than at $\alpha_\omega = 0°$. From the figure, it can be seen that the truck and the light trailer are more easily influenced by strong crosswind. In addition, compared to the bus and large trailer, the van is more sensitive to strong crosswind. Similar results for the failure probability of different vehicles running on dry road at $v_0 = 30$ km/h can be seen in Figure 7.10. However, attention should be paid to the fact that the failure probability of the car with light trailer at $v_0 = 30$ km/h is larger than that of the truck at same vehicle speed with $\alpha = 30°$, which differs from the result at $v_0 = 60$ km/h.

Moreover, in order to study the influence of road conditions on the crosswind stability of vehicles, the vehicle running on two different road surfaces (i.e. dry road with friction coefficient of 0.9 and wet road with friction coefficient of 0.5) is investigated. In comparison to the dry road, wet road can increase the failure probability of the vehicles. An example of the failure probability of the passenger car running on wet road with different vehicle speeds can be seen in Figure 7.11. Comparing Figure 7.11 with Figure 7.8, it can be found that on the same level of failure probability, the passenger car running on dry road can withstand a higher wind speed than that on wet road. Take $P_G = 1 \times 10^{-2}$ and $v_0 = 60$ km/h for example, the corresponding critical wind speed for the dry road and wet road are approximately 28 m/s and 21 m/s , respectively. Besides, in Figure 7.11 it can

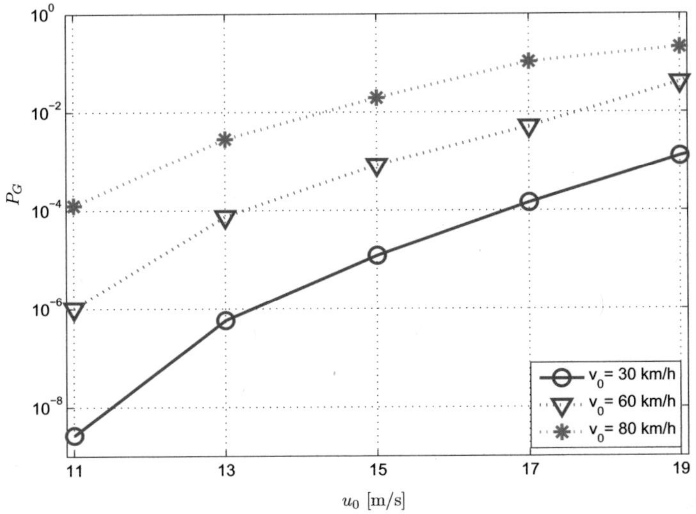

Figure 7.7: Failure probability for the truck running on dry road at different vehicle speeds

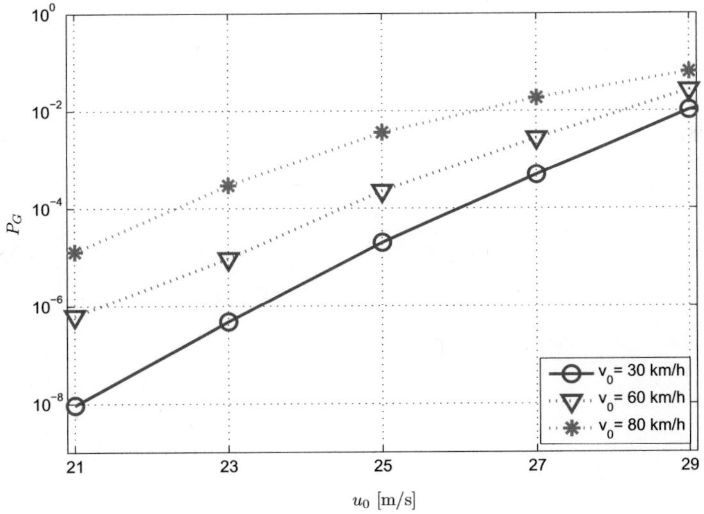

Figure 7.8: Failure probability for passenger car running on dry road with different vehicle speeds

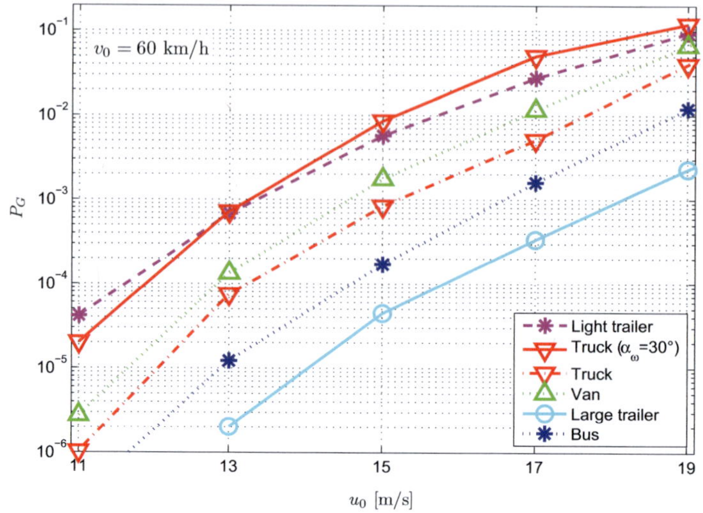

Figure 7.9: Failure probability of different vehicles running at $v_0 = 60$ km/h on dry road

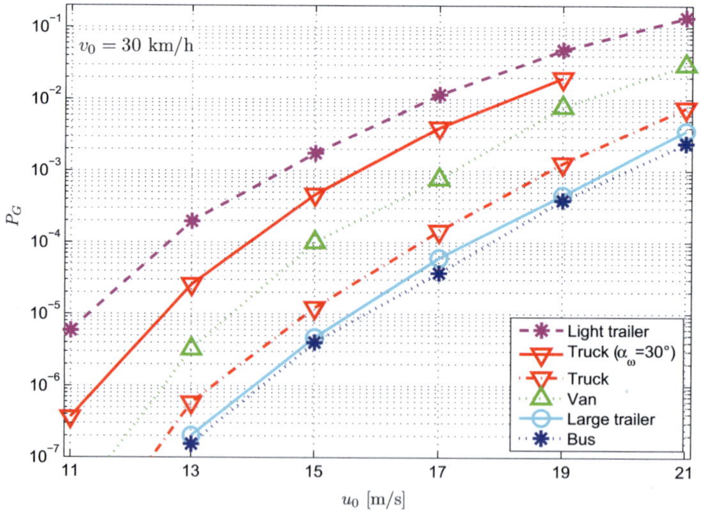

Figure 7.10: Failure probability of different vehicles running at $v_0 = 30$ km/h on dry road

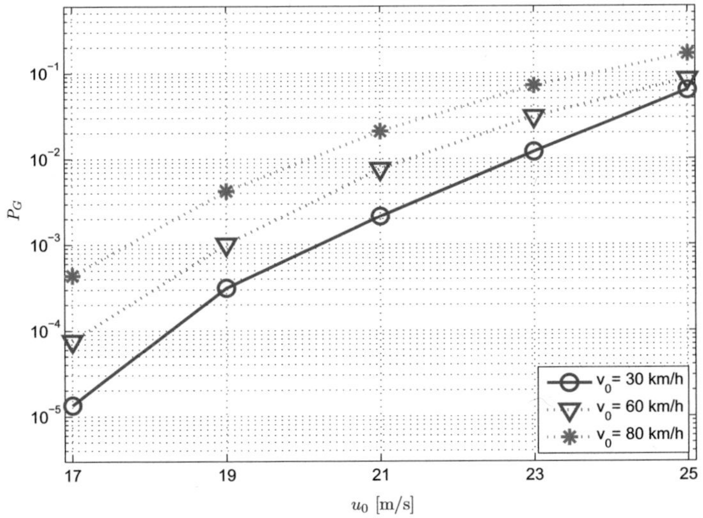

Figure 7.11: Failure probability of the passenger car running on wet road with different vehicle speeds

be also found that the vehicle speed still plays an important role in the crosswind stability of vehicles on wet road. Same as on dry road, larger vehicle speed will lead to a higher failure probability.

Figure 7.12 shows the results for the failure probabilities of other vehicles running on wet road. The vehicle with light trailer is still the most sensitive to strong crosswind. However, it has been noted that the large trailer on wet road has a higher failure risk than the truck and the bus. This is different from the conclusion obtained on dry road. The main reason may be due to the lower adhesion force of the large trailer on wet road. For the large trailer with relatively lighter weight (compared to the truck), the side slip is more easily to occur. On the other hand, the heavier weight of the truck leads to a larger adhesion force and consequently, the failure mode of the truck on wet road is overturning rather than side slip of the car with large trailer.

In Figure 7.13, the effect of road irregularities on the failure probability of the vehicle is displayed. It can be seen that there is visible but not so large increase in the failure probability when road irregularities are taken into account.

Furthermore, the influence of wind angles and nonstationary wind turbulence on crosswind stability of vehicles has also been investigated.

Figure 7.14 compares the failure probability of the truck running at different wind angles. As can be seen from the figure, the highest failure probability happens at $\alpha_\omega = 30°$, which implies that the most dangerous state of a moving vehicle under strong crosswind occurs not always under condition $\alpha_\omega = 0°$.

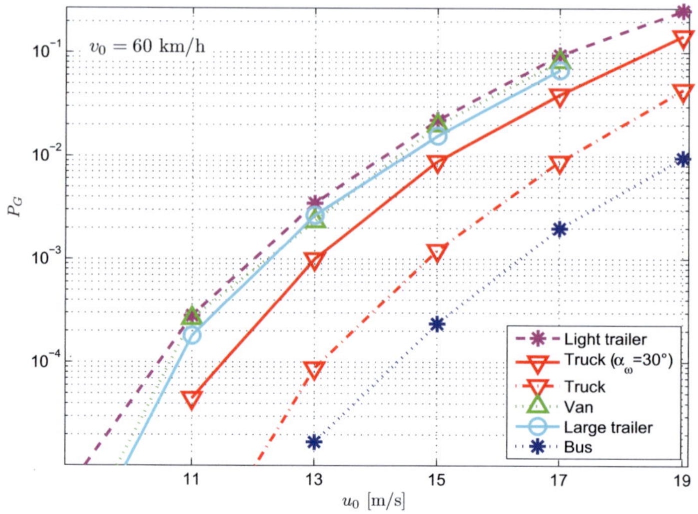

Figure 7.12: Failure probability of different vehicles running on wet road at $v_0 = 60$ km/h

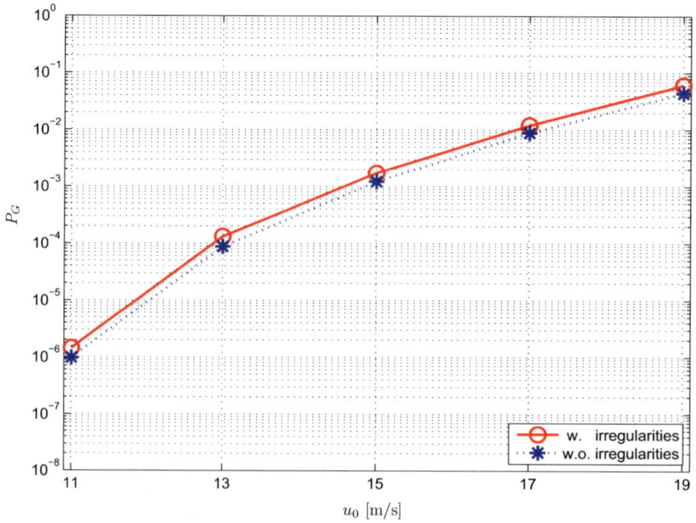

Figure 7.13: Influence of road irregularities on the failure probability of the vehicle (e.g. truck) running at $v_0 = 60$ km/h on wet road

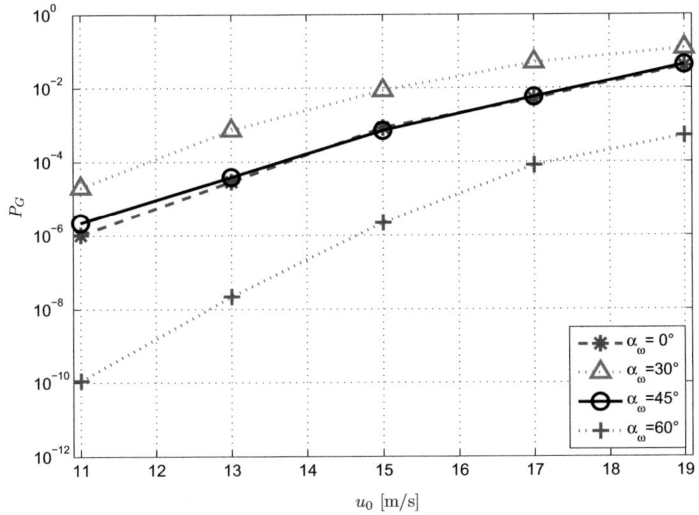

Figure 7.14: Failure probability of the truck running at $v_0 = 60$ km/h with different wind angles on dry road

As expected, when the wind angle becomes larger, e.g. $\alpha_\omega = 60°$, the failure probability will become lower. Figure 7.15 represents the failure probability of different vehicles running at $v_0 = 60$ km/h with $\alpha_\omega = 60°$ on dry road. Compared to the vehicles running at $v_0 = 60$ km/h with $\alpha_\omega = 0°$, the failure probability for a same wind speed decreases a lot.

Figure 7.16 displays the influence of nonstationary wind turbulence on the failure probability of the truck running at $v_0 = 60$ km/h. It once again proves that the nonstationary wind turbulence affects the crosswind stability of vehicles.

Another important parameter which has a great influence on crosswind stability is the vehicle mass. In general, vehicles with larger mass will have a smaller failure probability. An empty vehicle will be more easily influenced by strong crosswind. Take a van for example, when its mass changes from $m = 1900$ kg (empty vehicle) to $m = 2400$ kg (loaded vehicle), the failure probability will decrease a lot, see Figure 7.17. Moreover, Figure 7.18 compares the failure probability of the van running in strong crosswind with different vehicle heights. The van with lower height will have a smaller failure risk, which once again shows the influence of the vehicle shape on crosswind stability.

By now, it can be concluded that due to the difference of nonlinear aerodynamic coefficients as well as the different vehicle parameters (including the vehicle lengths, heights and masses), the influence of strong crosswinds on the crosswind stability of different vehicles in various scenarios are not always the same although there are still some common effects (e.g. larger vehicle speed or wind speed will give rise to a larger failure probability).

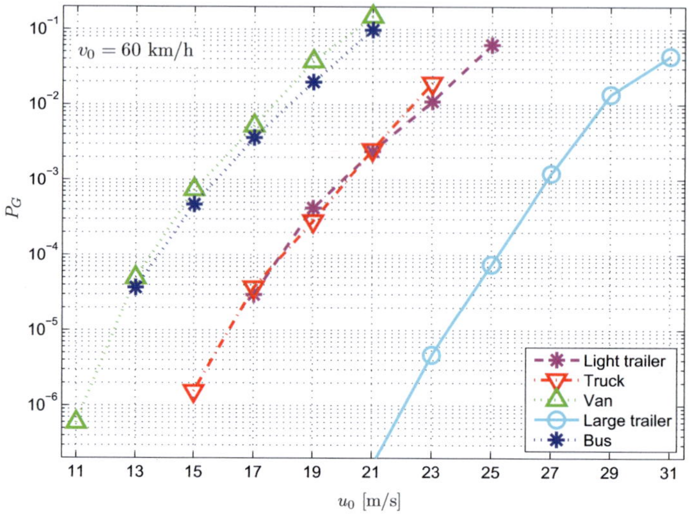

Figure 7.15: Failure probability of different vehicles running at $v_0 = 60$ km/h, $\alpha_\omega = 60°$ on dry road

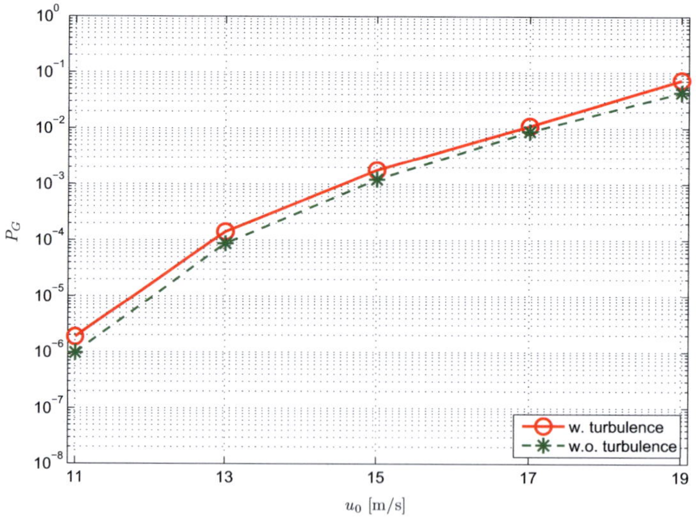

Figure 7.16: Influence of nonstationary wind turbulence on the failure probability of the truck running at $v_0 = 60$ km/h on wet road

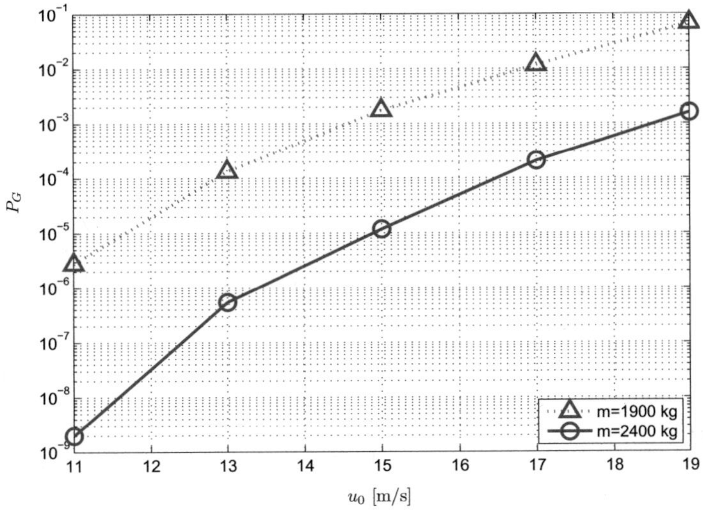

Figure 7.17: Influence of the vehicle mass on the failure probability

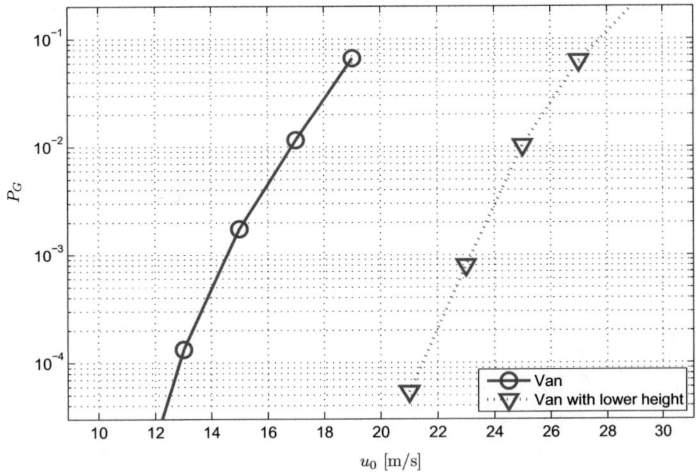

Figure 7.18: Influence of the vehicle height on the failure probability

Realistic assessments for the influence of strong crosswinds on different vehicles have to identify the specific conditions (e.g. on dry road or wet road, wind direction).

In summary, the proposed method gives a good description for the vehicle stability under strong crosswinds. It makes the quantitative comparison of the failure probability for road vehicles under various conditions possible. Based on the risk analysis, a vehicle speed limit for the safety of vehicles can be proposed.

7.4 Practical application: Wind warning system

As mentioned before, extreme wind events can significantly influence the safety and performance of road vehicles. Although the accidents caused by strong crosswinds occupy a relatively small proportion in comparison to other accidents, they may have a disproportionate contribution to the serious and fatal injuries. In order to reduce the damage caused by strong crosswind to the road vehicles, countermeasures such as reducing the wind speed by wind fences or enhancing the ability to resist to the wind have to be implemented. In practice, one possible way to prevent wind related accidents is to alert the drivers in advance or set restrictions (e.g. vehicle speed limit, route closure). In this regard, an investigation of wind warning system with signaling program for different vehicles is usually carried out.

The determination for vehicle speed limit (or critical wind speed) should take many aspects into consideration. On the one hand, the limit value can not be set too high, otherwise the warning system may not be able to ensure adequate safety. On the other hand, too conservative evaluation of the vehicle speed limit will make the warning system be activated too frequently, which may lead to a bad influence on traffic and cause complaints.

This section introduces a method for determining the vehicle speed limit (or critical wind speed) in practice based on the above risk analysis results. Before describing this method, critical wind speeds of vehicles for a wind warning system proposed by earlier study are discussed first.

7.4.1 Wind warning system by earlier study

In the past few decades, several studies have been conducted on the wind warning system. Baker [9] proposed a "two-level" system of control program for road vehicles based on a deterministic wind model. The considered types of vehicles include passenger cars, lorries, buses and trucks with trailers. For each type of the vehicle, the critical wind speeds with relation to the wind angles are given out. The proposed wind warning system contains the following two parts:

- At gust speeds above 17.5 m/s, the vehicle speed for high-sided vehicles will be restricted to about 10 m/s and a warning sign will be activated.

- When the gust speed is above 22.5 m/s, a closure of the whole traffic route is recommended.

Later, Baker [10] presented an improved vehicle model used in [7] by introducing a driver model. An risk analysis for the articulated lorry was carried out. Studies showed that the driver model, although very simplified, has a realistic influence on the crosswind. Besides, the study also showed that the value of critical wind speeds studied based on vehicle models without driver may be overestimated. Next, in [17], based on the analysis of wind-induced accidents of road vehicles that occurred in UK during the major storm in January, 1990, Baker pointed out that overturning accidents are the most common type of wind-induced accident and the traffic movement was suggested to be restricted at gust wind speed larger than 17 m/s to 20 m/s for all vehicles.

Recently, in [125], a signaling program for the wind warning system in the state of Oregon, USA has been introduced. The wind-warning system consists of two levels, i.e.

- A warning signal for the mean wind speed larger than 15.6 m/s for two minutes.
- A full closure for the route at mean wind speed larger than 35.8 m/s for two minutes.

Reference [143] reports the signaling program of the wind warning system in three different places of the USA based on the analysis of real accident data. The program is proposed to reduce the overturning risk of trucks running in strong crosswinds and can be described as follows:

- A warning info will be displayed at gust speed above 15.6 m/s on dry road and 13.4 m/s on icy road, respectively.
- A closure of the route for cars with trailers at gust speed above 20.1 m/s on dry road or 17.9 m/s on icy road will be implemented. For trucks or cars with heavier trailers, the limit values of gust speed on dry and icy roads are 26.8 m/s and 20.1 m/s, respectively.
- The route will be completely closed when a gust speed achieves 31.3 m/s on dry road and 24.6 m/s on icy road.

In reference [215] and [61], an investigation of the intelligent transportation system (ITS) for improving the truck safety under strong crosswinds has been carried out. Based on the statistical analysis of past wind-induced accidents of trucks, a similar three-level signaling program for trucks has been proposed, i.e.

- A warning info will be given out when the mean wind speed is higher than 13.4 m/s.
- At mean wind speed above 17.9 m/s, the route will be closed for high-light trucks.
- At mean wind speed above 20.1 m/s, the route will be closed for all large trucks.

By verifying the frequency of the suggested threshold on the past wind data in the State of Wyoming, USA, about 88% of the wind speeds are below the value of 13.4 m/s which ensures that the warning will not be activated too frequently.

7.4.2 Wind warning system based on reliability analysis

In general, the above conclusions provide a good reference for the design of wind warning systems. However, the simulation models for the previous study are usually based on deterministic wind models, which in reality is not correct.

In order to analyze the critical wind speed in a more realistic way, the stochastic wind model has to be utilized and the results of risk assessment have to be referred. Following this, a method for determining the critical wind speed based on reliability analysis will be introduced.

Firstly, due to the fact that the failure probability of the passenger car is significantly lower compared to that of the other vehicles, cf. Figure 7.8 and Figure 7.9, it is advisable to take the trucks, vans, buses and cars with light or large trailers as one group and the passenger cars as the other group.

Then, before determining the critical wind speed for each type of vehicle in a specified wind scenario (e.g. dry road with perpendicular wind direction), an appropriate level for the conditional failure probability has to be defined. This can be done based on the statistical analysis of actual wind-induced accidents or the reference values given by the literature (cf. the example given in the following).

After the appropriate level for the conditional failure probability is found, a safety level based on the conditional failure probability can be proposed.

Finally, the recommended critical wind speeds for a certain vehicle group in a specified scenario can be obtained from the intersections of the probabilistic characteristic curve with the safety level lines.

For example, assume that the cars with light trailers are the most prone to have wind-induced accidents at a certain place. And the statistics show that frequent accidents for such type of vehicles mainly occurred at $u_0 \approx 18.5$ m/s and $v_0 \approx 60$ km/h on wet road. Then based on the failure probability results, the corresponding probability level for the accident point would be approximately around $P_{f0} = 2 \times 10^{-1}$, cf. Figure 7.19. According to this level, a possible safety level below P_{f0} could be proposed and the corresponding wind speed can be regarded as critical wind speed. For instance, if a probability level of 2×10^{-2} has been chosen as the safety level, then the corresponding mean wind speed for the recommended safety level will be $u_0 = 15$ m/s, which means that to reduce the failure risk, the speed of the car with light trailer (including other vehicles classified into the same group as the car with light trailer) should be restricted to 60 km/h when the mean wind speed achieves 15 m/s.

The safety level obtained above could be adjusted in future step by step (i.e. increasing or decreasing the level). In practice, to be safe, a safety margin, namely, the conservative safety level can be applied at first. The conservative safety level, which is relatively lower than the safety level, can be determined according to experience or other factors (e.g. requirement of the safety margin) [143] [61] and then increased little by little in practice

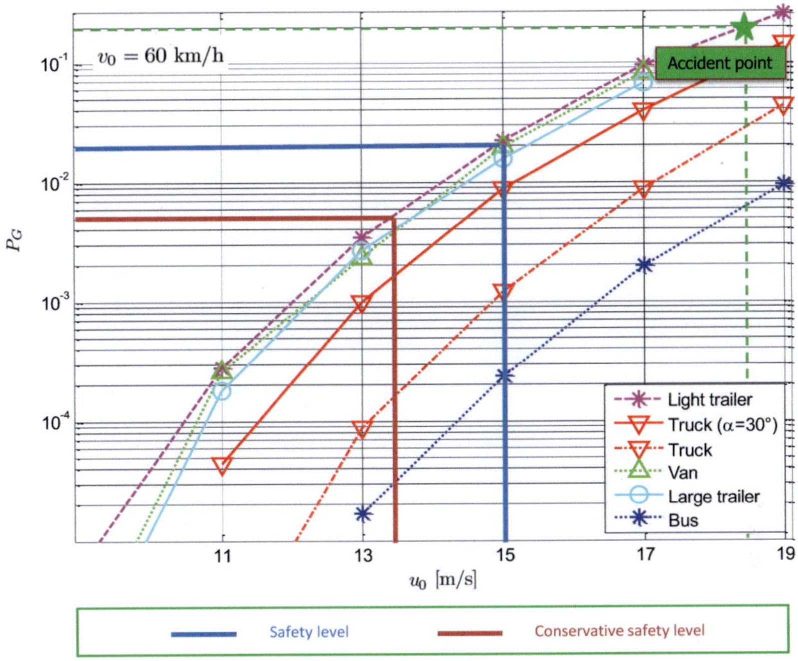

Figure 7.19: Threshold value of mean wind speed for road vehicles on wet road, $\alpha_\omega = 0°$

until a suitable value is achieved. An example of the conservative safety level at $P_G = 5 \times 10^{-3}$, which leads to a critical wind speed $u_0 = 13.5$m/s can be seen in Figure 7.19.

Similarly, for passenger cars, the critical wind speed at the conservative safety level $P_G = 5 \times 10^{-3}$ at vehicle speed $v_0 = 60$ km/h on wet road is $u_0 \approx 20.5$ m/s, cf. Figure 7.11.

The critical wind speeds obtained above is proposed for vehicles running on wet road. For vehicles running in other scenarios (e.g. dry road or wet road with other wind angles), the same method could be applied. It should be noted again that the reference point (i.e. the accident point in Figure 7.19) should be determined based on both theoretical analysis and practice. A further verification and improvement for such a wind warning system can be carried out according to the future wind-induced accident data, so that a final, more robust wind warning system can be obtained.

8 Summary and Conclusions

8.1 Main results and contributions

The general aim of the work presented in this thesis is to study the nonstationary crosswind stability of ground vehicles and thus to establish a realistic assessment for the vehicle running under strong crosswind. In this regard, following contents have been mainly investigated and proposed:

1. The nonstationary characteristics of the wind speed are studied. Based on the analysis of nonstationary crosswind, a stochastic gust model with nonstationary wind turbulence based on NONAR method has been proposed and applied for the analysis of crosswind stability of vehicles.

2. Considering the actual effect of turbulent wind acting on a vehicle, the wind excitation of a moving vehicle with a spatial dimension has been analyzed and the admittance function has been utilized to calculate the nonstationary wind excitation. In addition, wind forces induced by vortex shedding and buffeting are discussed and the Hilbert-Huang transform (HHT) is applied to analyze the nonstationary wind forces.

3. Crosswind stability of railway vehicles running on both straight and curved tracks has been investigated. Based on the nonstationary crosswind process, a vehicle running in a nonstationary vehicle state (acceleration/deceleration) has been modeled and simulated. Besides, risk analysis has been carried out for the vehicle running on curved track with various design and track parameters. Based on sensitivity analysis the most important parameters for crosswind stability have been identified.

4. Road vehicles are classified into several categories. For each vehicle type, a worst case vehicle model and the corresponding aerodynamic coefficients have been identified. In addition, to be more realistic, the wind data measured at the exposed site are used to calibrate the nonstationary gust model. The influences of the stochastic crosswind on different vehicles in various scenarios have been analyzed and compared.

5. The presented simulation and results allow to base a signaling program for the actual traffic influence on a risk level and the effect of various countermeasures such as wind fences (wind speed reduction) and vehicle speed reduction can be judged on an objective basis.

In Chapter 2, fundamental characteristics of the wind speed are presented. Based on the analysis of natural wind speed, the probability density functions (PDFs) of the wind speed are discussed and the Weibull PDF is used to fit the actual PDF of measured wind data. Results show that Weibull PDF fits well to the PDF of measured wind data. In addition, this chapter also studies several different methods for the simulation of wind turbulence, namely the traditional PSD method, the wavelet method and the NONAR method. The traditional PSD method cannot be utilized directly to simulate the nonstationary turbulence. Compared to the wavelet method, the NONAR method based on ARMA model is considered to be computationally more efficient. Therefore, it has been applied in this work. At the end of Chapter 2, the stochastic gust model with nonstationary wind turbulence as well as the distribution of the normalized gust amplitude and duration are described. The gust characteristic based on coherence function changes with both mean wind and vehicle speeds. Moreover, the PDF of normalized gust amplitude and duration are fitted by a half-Gaussian function and a log-normal function, respectively.

Following Chapter 2, the wind excitation caused by nonstationary crosswind acting on a moving vehicle is studied in Chapter 3. Based on the analysis of a vehicle with spatial dimension excited by nonstationary crosswind, the equivalent wind spectrum which includes an aerodynamic admittance function has been proposed to compute the nonstationary wind excitation. For a moving vehicle, the shape of the equivalent spectrum changes with the vehicle speed. Besides, it is considered that the wind excitation acting on a vehicle is usually caused by vortex shedding and buffeting. In order to study the time-frequency characteristic of the nonstationary wind, the Hilbert-Huang transform (HHT) has been applied for the analysis. By HHT, the trend and time-frequency characteristics of the nonstationary wind force can be well seen. At every instant of time, there are several different frequencies involved in the wind speed. Increasing vehicle speed relative to wind speed will move the energy content of the wind spectrum to a higher frequency range.

Chapter 4 details the mathematical model and physical scenario for both road and railway vehicles running under the excitation of crosswinds.

On the one hand, road vehicles are classified into several categories and modeled in Matlab/Simulink as a combination of rigid bodies with lumped masses, suspensions and wheels. Except for the car with trailer, an unified two-axle and four-wheel vehicle model has been adopted. For the car with trailer, a nonlinear double-track model with two separate parts, i.e. the car and the trailer, has been studied. Both the driver and road condition are taken into account. For each vehicle type, the corresponding vehicle parameters and aerodynamic forces are identified. Studies show that strong crosswinds have an influence on the stability of vehicles. When a crosswind comes, the lateral displacement of the vehicle and the yaw angle as well as the vertical wheel force will change. Extreme strong winds may even lead the vehicle to an uncontrollable state. Besides, the driver has also an influence on the crosswind stability. As can be imagined, a driver with experience can control the vehicle much better. Moreover, the crosswind stability can also be affected by the road irregularity and wind turbulence.

The railway vehicle model is built in the commercial software ADAMS/Rail as a classical double-decker cabin car. Nonlinear characteristics of the suspension are simulated by springs and dampers. Besides, the bump stops, which have great influence on the lateral behavior of the vehicle, have also been taken into account. Wheel/rail contact forces are calculated by Kalker's FASTSIM algorithm since it provides a good compromise between simulation accuracy and simulation speed. As one important part of the wind-vehicle system, track geometries have a great influence on crosswind stability. In addition, the unbalanced lateral acceleration of the vehicle also affects the crosswind stability. For a given curved track, the permissible vehicle speed changes with the permissible cant deficiency $h_{e,p}$ or the permissible lateral acceleration $a_{y,p}$.

Chapter 5 introduces a probabilistic method for evaluating the crosswind stability of vehicles with respect to a single gust event. It takes uncertain parameters (such as aerodynamic coefficients, gust amplitude, gust duration and so on) into account and yields the failure probabilities instead of traditional critical results based on reliability analysis. To calculate the failure probability, failure criteria and corresponding performance functions are defined. The main failure mode for railway vehicles caused by crosswind is overturning. For road vehicles, the failure criteria become more complicated due to the complex traffic situation. It has been considered that there are mainly three failure modes for road vehicles, namely, side slip, rotation and overturning. An accident is said to happen if any one of the three failure modes occurs. Calculations of the failure probability are mainly based on FORM and LS methods. In addition, failure probability of the vehicle running in a relatively long time interval has also been studied. To be efficient, the maximum-value method and peak-over-threshold (POT) method are proposed to calculate the failure probability for such a case.

Chapter 6 studies the crosswind stability of railway vehicles running under the excitation of nonstationary crosswind on both straight and curved tracks. Besides, the failure probability of a vehicle running in a relatively long journey or passing through an artificial storm with acceleration has also been computed. Studies show that the vehicle speed with nonstationary wind excitation plays an important role in the crosswind stability of vehicles. The failure probability increases when the vehicle accelerates. For a vehicle running on curved track, the scenario becomes more complicated. It has been noted that a vehicle running on curved track with certain vehicle speeds can undergo unbalanced lateral accelerations. In this respect, the vehicle running on curved track with outer-side wind excitation has to be distinguished from that with inner-side wind excitation. Results show that the wind speed, wind direction as well as the wind turbulence have influences on the crosswind stability. Winds coming from the inner side (centrifugal direction) will increase the failure probability while winds coming from the outer side (centripetal direction) may decrease the failure probability. For a given wind direction, the main influence factor for the crosswind stability of vehicles running on curved track is the unbalanced lateral acceleration. Next, at the end of Chapter 6, sensitivity analysis has been carried out. Studies show that the gust amplitude A, the aerodynamic coefficient C_{mx} and the gust duration T are the most influential parameters on the crosswind stability of railway vehicles.

In Chapter 7, a risk analysis has been carried out for the crosswind stability of road vehicles. As mentioned before, the road vehicles are classified into several types. For each vehicle type, the worst case vehicle model and corresponding aerodynamic coefficients are identified. In order to be more realistic, wind data measured at exposed locations are used to calibrate the nonstationary gust model. Studies show that strong crosswind has a great influence on road vehicles. For different vehicle types, the influences are not always the same. However, there are still some general characteristics in common, e.g. the failure probability will become larger with higher vehicle and wind speeds. In addition, the influence of wind angle on the crosswind stability of vehicles has also been investigated. Moreover, this chapter also studies the crosswind stability of vehicles running on wet road. In general, wet road condition will move the vehicle to a more dangerous state. Based on the results of risk analysis for vehicles in various scenarios, the vehicle speed limit for a certain vehicle type under different conditions (e.g. on dry road or wet road) can be determined. An example for the application of determining the vehicle speed limit for a wind warning system based on risk level has been presented in the last section of this chapter.

8.2 Recommendations for further studies

This work has studied the crosswind stability of railway and road vehicles under non-stationary crosswind. In light of the present models and conclusions, following work can be suggested for further research so that a more accurate assessment for the crosswind stability of vehicles can be obtained.

To begin with, nonlinear aerodynamic coefficients can affect the evaluation of the wind excitation greatly. To get an accurate assessment for the failure probability of ground vehicles, it is necessary to identify the aerodynamic coefficients for each vehicle type more accurately. In view of this, more experiments on studying the aerodynamic coefficients have to be carried out. In the experiment, the vehicle geometric features, corresponding wind scenarios as well as the vehicle states have to be clarified since these parameters may lead to quite different results.

Besides, in this work, the distributed wind excitation is computed approximately as concentrated forces and moments by taking an aerodynamic admittance function into account. Further studies and more detailed analyses can be carried out on the relationship between the distributed wind force and the concentrated force. On this point, numerical simulations as well as full-scale and wind-tunnel tests are recommended.

In addition, more research work could be suggested on the instantaneous wind force caused by nonstationary physical scenarios (e.g. transient wind, moving vehicle with complex driving scenario). A comparison between the transient wind force and the quasi-steady wind force for the vehicle in time domain could be investigated in a further step.

Finally, studies for the influence of different advanced devices (e.g. active suspensions) on the crosswind stability of road vehicles would also be of interest. More studies could be carried out on the vehicle parameters so that a more robust design of the vehicle can be proposed.

Appendix

A Parameters for different vehicle models

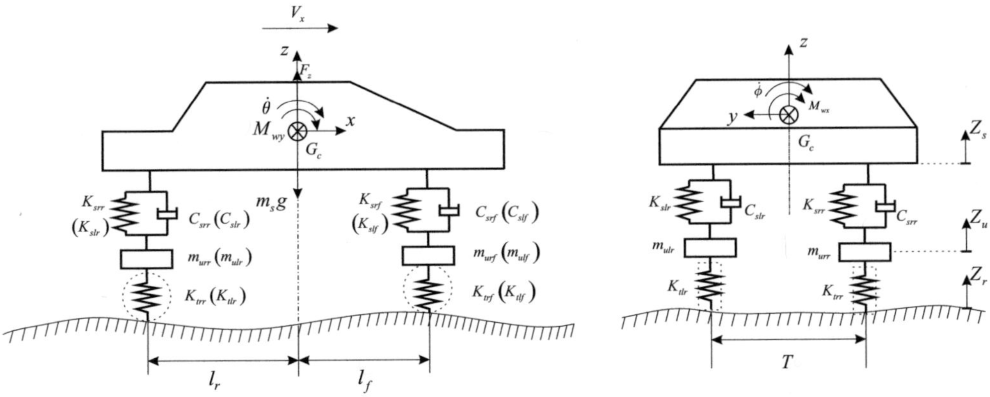

Figure A.1: Car model

Figure A.1 shows a model for the passenger car. The basic principle for modeling the car is the same as other vehicles with two axes and four wheels (e.g. a truck model in Chapter 4). Main differences for modeling these vehicles are in vehicle parameters and corresponding aerodynamic coefficients.

In the following, main parameters for different vehicles used in this dissertation are listed. They are mainly obtained from the published papers or based on experience, cf. [32] [110] [28] [208] [166] and [9]

Table A.1: Main parameters for the bus used in the numerical simulation

m	Vehicle mass	12000 kg
m_s	Sprung mass	11000 kg
l_f	Distance between c.g. and the front axle	6.0 m
l_r	Distance between c.g. and the rear axle	3.0 m
L	Vehicle length	12 m
H	Vehicle height	3 m
I_{xx}	Rolling moment of inertia	1352 kgm^2
I_{yy}	Pitching moment of inertia	5483 kgm^2
I_{zz}	Yawing moment of inertia	60000 kgm^2
K_{sf}	Upper vertical spring stiffness	400000 N/m
K_{sr}	Lower vertical spring stiffness	400000 N/m
C_f	Upper vertical damping stiffness	20000 Ns/m
C_r	Lower vertical damping stiffness	20000 Ns/m

Table A.2: Main parameters for the van used in the numerical simulation

m	Vehicle mass	1900 kg
m_s	Sprung mass	1800 kg
l_f	Distance between c.g. and the front axle	1.5 m
l_r	Distance between c.g. and the rear axle	3.0 m
L	Vehicle length	7 m
H	Vehicle height	3 m
I_{xx}	Rolling moment of inertia	1352 kgm^2
I_{yy}	Pitching moment of inertia	5483 kgm^2
I_{zz}	Yawing moment of inertia	20000 kgm^2
K_{sf}	Upper vertical spring stiffness	400000 N/m
K_{sr}	Lower vertical spring stiffness	400000 N/m
C_f	Upper vertical damping stiffness	20000 Ns/m
C_r	Lower vertical damping stiffness	20000 Ns/m

Table A.3: Main parameters for the passenger car used in the numerical simulation

m	Vehicle mass	1500 kg
m_s	Sprung mass	1300 kg
l_f	Distance between c.g. and the front axle	1.07 m
l_r	Distance between c.g. and the rear axle	1.47 m
L	Vehicle length	5 m
H	Vehicle height	1.5 m
I_{xx}	Rolling moment of inertia	498.9 kgm^2
I_{yy}	Pitching moment of inertia	500 kgm^2
I_{zz}	Yawing moment of inertia	2700 kgm^2
K_s	Upper vertical spring stiffness	66185.8 N/m
C_s	Upper vertical damping stiffness	3511.6 Ns/m

Table A.4: Main parameters for the car with trailer used in the numerical simulation

m	Vehicle total mass (powered unit)	1500 kg
m_s	Vehicle sprung mass (powered unit)	1300 kg
l_f	Distance between c.g. and the front axle (powered unit)	1.4 m
l_r	Distance between c.g. and the rear axle (powered unit)	1.7 m
L	Vehicle length (powered unit)	5 m
H	Vehicle height (powered unit)	1.5 m
I_{xx}	Rolling moment of inertia (powered unit)(powered unit)	498.9 kgm^2
I_{yy}	Pitching moment of inertia (powered unit)	500 kgm^2
I_{zz}	Yawing moment of inertia (powered unit)	6784 kgm^2
m_t	Vehicle total mass (trailer unit)	1500 kg
m_{ts}	Vehicle sprung mass (trailer unit)	1300 kg
l_{ft}	Distance between c.g. and the front axle (trailer unit)	1.25 m
l_{rt}	Distance between c.g. and the rear axle (trailer unit)	1.55 m
L_t	Vehicle length (trailer unit)	5 m
H_t	Vehicle height (trailer unit)	2 m
I_{xxt}	Rolling moment of inertia (trailer unit)	498.9 kgm^2
I_{yyt}	Pitching moment of inertia (trailer unit)	500 kgm^2
I_{zzt}	Yawing moment of inertia (trailer unit)	3441 kgm^2

Table A.5: Main parameters for the car with light trailer used in the numerical simulation

m	Vehicle total mass (powered unit)	1500 kg
m_s	Vehicle sprung mass (powered unit)	1300 kg
l_f	Distance between c.g. and the front axle (powered unit)	1.4 m
l_r	Distance between c.g. and the rear axle (powered unit)	1.7 m
L	Vehicle length (powered unit)	5 m
H	Vehicle height (powered unit)	1.5 m
I_{xx}	Rolling moment of inertia (powered unit)(powered unit)	498.9 kgm^2
I_{yy}	Pitching moment of inertia (powered unit)	500 kgm^2
I_{zz}	Yawing moment of inertia (powered unit)	6784 kgm^2
m_t	Vehicle total mass (trailer unit)	150 kg
m_{ts}	Vehicle sprung mass (trailer unit)	130 kg
l_{ft}	Distance between c.g. and the front axle (trailer unit)	0.4 m
l_{rt}	Distance between c.g. and the rear axle (trailer unit)	0.4 m
L_t	Vehicle length (trailer unit)	1.45 m
H_t	Vehicle height (trailer unit)	1.2 m
I_{xxt}	Rolling moment of inertia (trailer unit)	49.89 kgm^2
I_{yyt}	Pitching moment of inertia (trailer unit)	50 kgm^2
I_{zzt}	Yawing moment of inertia (trailer unit)	582.36 kgm^2

B Aerodynamic coefficients

Typical aerodynamic coefficients for the railway vehicle please refer to [202] [33] and [15]. In the following, the aerodynamic coefficients for different road vehicles are presented.

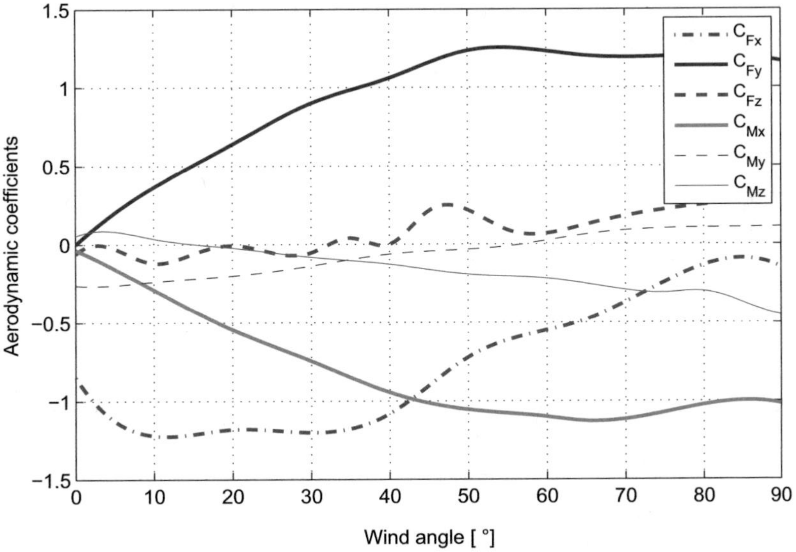

Figure B.1: Aerodynamic coefficients for the truck [38]

C_{Fx}: Aerodynamic coefficient for the longitudinal wind force

C_{Fy}: Aerodynamic coefficient for the lateral wind force

C_{Fz}: Aerodynamic coefficient for the vertical wind force

C_{Mx}: Aerodynamic coefficient for the roll moment

C_{My}: Aerodynamic coefficient for the pitch moment

C_{Mz}: Aerodynamic coefficient for the yaw moment

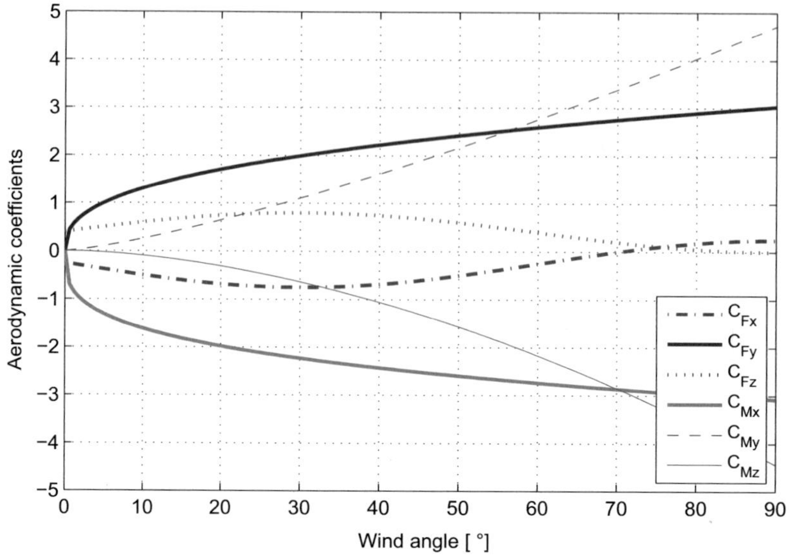

Figure B.2: Aerodynamic coefficients for the passenger car [9]

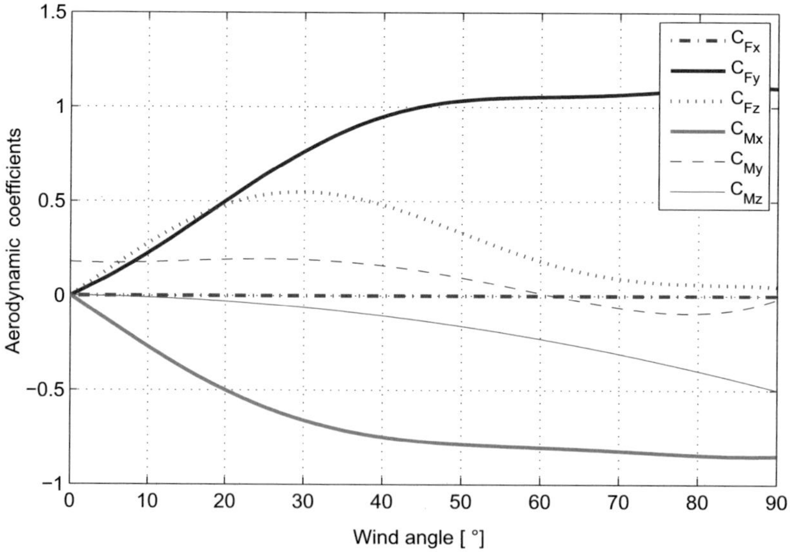

Figure B.3: Aerodynamic coefficients for the bus [113]

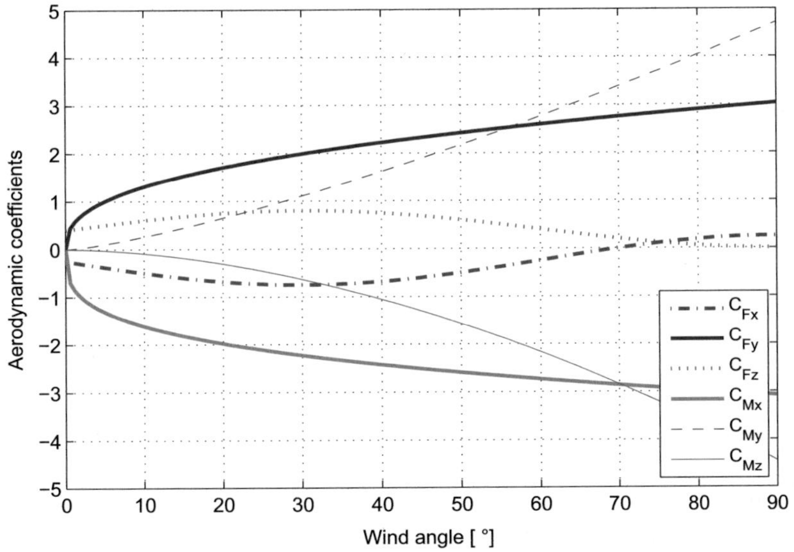

Figure B.4: Aerodynamic coefficients for the van [9]

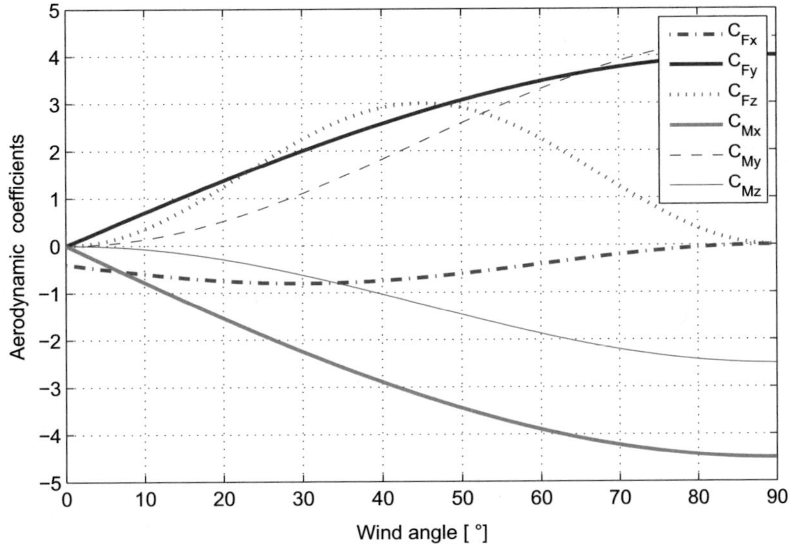

Figure B.5: Aerodynamic coefficients for the trailer [17]

List of Figures

List of Tables

Bibliography

[1] R. Adhikari and H. Yamaguchi. A study on the nonstationarity in wind and wind-induced response of tall buildings for adaptive active control. *Journal of wind engineering and industrial aerodynamics*, 72:213–224, 1997.

[2] R. W. Allen, R. E. Magdaleno, T. J. Rosenthal, D. H. Klyde, and J. R. Hogue. Tire modeling requirements for vehicle dynamics simulation. *SP-Society of Automotive Engineers*, (1074):95–115, 1995.

[3] K. J. Åström and T. Hägglund. Pid controllers: theory, design, and tuning. *Instrument Society of America, Research Triangle Park, NC*, 1995.

[4] K. J. Åström, T. Hägglund, C. C. Hang, and W. K. Ho. Automatic tuning and adaptation for pid controllers-a survey. *Control Engineering Practice*, 1(4):699–714, 1993.

[5] S. K. Au and J. L. Beck. Estimation of small failure probabilities in high dimensions by subset simulation. *Probabilistic Engineering Mechanics*, 16(4):263–277, 2001.

[6] S. K. Au, J. Ching, and J. L. Beck. Application of subset simulation methods to reliability benchmark problems. *Structural Safety*, 29(3):183–193, 2007.

[7] C. Baker. A simplified analysis of various types of wind-induced road vehicle accidents. *Journal of Wind Engineering and Industrial Aerodynamics*, 22(1):69–85, 1986.

[8] C. Baker. Train aerodynamic forces and moments from moving model experiments. *Journal of Wind Engineering and Industrial Aerodynamics*, 24(3):227–251, 1986.

[9] C. Baker. Measures to control vehicle movement at exposed sites during windy periods. *Journal of Wind Engineering and Industrial Aerodynamics*, 25(2):151–161, 1987.

[10] C. Baker. High sided articulated road vehicles in strong cross winds. *Journal of Wind Engineering and Industrial Aerodynamics*, 31(1):67–85, 1988.

[11] C. Baker. Ground vehicles in high cross winds part 2: unsteady aerodynamic forces. *Journal of fluids and structures*, 5(1):91–111, 1991.

[12] C. Baker. The effect of unsteady crosswind forces on train dynamic behaviour. In *Proceedings of the Fifth European and African Conference on Wind Engineering, Florence*, 2009.

[13] C. Baker. The flow around high speed trains. *Journal of Wind Engineering and Industrial Aerodynamics*, 98(6):277–298, 2010.

[14] C. Baker. The simulation of unsteady aerodynamic cross wind forces on trains. *Journal of Wind Engineering and Industrial Aerodynamics*, 98(2):88 – 99, 2010.

[15] C. Baker, F. Cheli, A. Orellano, N. Paradot, C. Proppe, and D. Rocchi. Cross-wind effects on road and rail vehicles. *Vehicle system dynamics*, 47(8):983–1022, 2009.

[16] C. Baker, J. Jones, F. Lopez-Calleja, and J. Munday. Measurements of the cross wind forces on trains. *Journal of Wind Engineering and Industrial Aerodynamics*, 92(7):547–563, 2004.

[17] C. Baker and S. Reynolds. Wind-induced accidents of road vehicles. *Accident Analysis & Prevention*, 24(6):559–575, 1992.

[18] C. Baker and M. Sterling. Aerodynamic forces on multiple unit trains in cross winds. In *BBAA VI International Colloquium on: Bluff Bodies Aerodynamics & Applications, Milano, Italy.*, 2008.

[19] E. Bakker, L. Nyborg, and H. B. Pacejka. Tire modeling for use in vehicle studies. *SAE paper*, (870421), 1987.

[20] P. W. Bearman. An investigation of the forces on flat plates in turbulent flow. *National Physical Laboratory, Aerodynamics Report 1296*, 1969.

[21] J. Bec. Influence of wind spectrum formula choice on footbridge response. *The Fifth International Symposium on Computational Wind Engineering (CWE2010), Chapel Hill, North Carolina, USA*, pages 23–27, 2010.

[22] D. Belušić and Z. B. Klaić. Estimation of bora wind gusts using a limited area model. *Tellus A*, 56(4):296–307, 2004.

[23] R. E. Benestad. Record-values, nonstationarity tests and extreme value distributions. *Global and Planetary Change*, 44(1):11–26, 2004.

[24] H. Bergström. A statistical analysis of gust characteristics. *Boundary-Layer Meteorology*, 39(1-2):153–173, 1987.

[25] J. E. Bernard and C. L. Clover. Tire modeling for low-speed and high-speed calculations. *SAE transactions*, 104(6):474–483, 1995.

[26] W. Bierbooms and P.W. Cheng. Stochastic gust model for design calculations of wind turbines. *Journal of Wind Engineering and Industrial Aerodynamics*, 90(11):1237–1251, 2002.

[27] W. Bierbooms, J. B. Dragt, and H. Cleijne. Verification of the mean shape of extreme gusts. *Wind Energy*, 2(3):137–150, 1999.

[28] B. L. Boada, M. Boada, and V. Diaz. Fuzzy-logic applied to yaw moment control for vehicle stability. *Vehicle System Dynamics*, 43(10):753–770, 2005.

[29] M. Bocciolone, F. Cheli, R. Corradi, S. Muggiasca, and G. Tomasini. Crosswind action on rail vehicles: wind tunnel experimental analyses. *Journal of Wind Engineering and Industrial Aerodynamics*, 96(5):584–610, 2008.

[30] E. Branlard. Wind energy: On the statistics of gusts and their propagation through a wind farm. Technical report, 2009.

[31] O. Brasseur. Development and application of a physical approach to estimating wind gusts. *Monthly Weather Review*, 129(1):5–25, 2001.

[32] C. Cai and S. Chen. Framework of vehicle–bridge–wind dynamic analysis. *Journal of Wind Engineering and Industrial Aerodynamics*, 92(7):579–607, 2004.

[33] A. Carrarini. *Reliability based analysis of the crosswind stability of railway vehicles.* PhD thesis, Technischen Universität Berlin, 2006.

[34] A. Carrarini. Reliability based analysis of the crosswind stability of railway vehicles. *Journal of Wind Engineering and Industrial Aerodynamics*, 95(7):493–509, 2007.

[35] D. E. Cartwright and M. S. Longuet-Higgins. The statistical distribution of the maxima of a random function. *Proceedings of the Royal Society of London. Series A. Mathematical and Physical Sciences*, 237(1209):212–232, 1956.

[36] M. T. Chay, F. Albermani, and R. Wilson. Numerical and analytical simulation of downburst wind loads. *Engineering Structures*, 28(2):240–254, 2006.

[37] F. Cheli, R. Corradi, D. Rocchi, G. Tomasini, and E. Maestrini. Wind tunnel tests on train scale models to investigate the effect of infrastructure scenario. *Journal of Wind Engineering and Industrial Aerodynamics*, 98(6):353–362, 2010.

[38] F. Cheli, R. Corradi, E. Sabbioni, and G. Tomasini. Wind tunnel tests on heavy road vehicles: Cross wind induced loads–part 1. *Journal of Wind Engineering and Industrial Aerodynamics*, 99(10):1000–1010, 2011.

[39] F. Cheli, S. Giappino, L. Rosa, G. Tomasini, and M. Villani. Experimental study on the aerodynamic forces on railway vehicles in presence of turbulence. *Journal of Wind Engineering and Industrial Aerodynamics*, 123:311–316, 2013.

[40] F. Cheli, F. Ripamonti, E. Sabbioni, and G. Tomasini. Wind tunnel tests on heavy road vehicles: Cross wind induced loads–part 2. *Journal of Wind Engineering and Industrial Aerodynamics*, 99(10):1011–1024, 2011.

[41] J. Chen, Michael C. Hui, and Y. Xu. A comparative study of stationary and non-stationary wind models using field measurements. *Boundary-layer meteorology*, 122(1):105–121, 2007.

[42] L. Chen and C. Letchford. Proper orthogonal decomposition of two vertical profiles of full-scale nonstationary downburst wind speeds. *Journal of Wind Engineering and Industrial Aerodynamics*, 93(3):187–216, 2005.

[43] L. Chen and C. W. Letchford. A deterministic–stochastic hybrid model of downbursts and its impact on a cantilevered structure. *Engineering structures*, 26(5):619–629, 2004.

[44] L. Chen and C. W. Letchford. Multi-scale correlation analyses of two lateral profiles of full-scale downburst wind speeds. *Journal of wind engineering and industrial aerodynamics*, 94(9):675–696, 2006.

[45] S. Chen and C. Cai. Accident assessment of vehicles on long-span bridges in windy environments. *Journal of Wind Engineering and Industrial Aerodynamics*, 92(12):991–1024, 2004.

[46] W. Chen, S. Wu, and Y. Zhang. Aerodynamic characteristics of high speed trains under cross wind conditions. In *Recent Progresses in Fluid Dynamics Research: Proceeding of the Sixth International Conference on Fluid Mechanics*, volume 1376, pages 181–183. AIP Publishing, 2011.

[47] X. Chen. Analysis of alongwind tall building response to transient nonstationary winds. *Journal of structural engineering*, 134(5):782–791, 2008.

[48] M. S. Cheung and Y. B. Chan. Vehicle-wind-long span bridge interaction and its effect on speed limit and vehicle stability. In *Proc., Joint Int. Conf. on Computing and Decision making in Civil and Building Engineering.* ICCCBE Montreal, Canada, 2006.

[49] T. W. Chiu. A two-dimensional second-order vortex panel method for the flow in a cross-wind over a train and other two-dimensional bluff bodies. *Journal of Wind Engineering and Industrial Aerodynamics*, 37(1):43–64, 1991.

[50] T. W. Chiu. Prediction of the aerodynamic loads on a railway train in a cross-wind at large yaw angles using an integrated two-and three-dimensional source/vortex panel method. *Journal of wind engineering and industrial aerodynamics*, 57(1):19–39, 1995.

[51] T. W. Chiu and L. C. Squire. An experimental study of the flow over a train in a crosswind at large yaw angles up to 90. *Journal of Wind Engineering and Industrial Aerodynamics*, 45(1):47–74, 1992.

[52] M. Clobes and A. Willecke. On the numerical simulation of gust and vortex excitation of guyed masts. *IASS Working Group No.4 - Masts and Towers 24th Meeting, Helsinki*, Sep., 2009.

[53] M. Clobes, A. Willecke, and U. Peil. A refined analysis of guyed masts in turbulent wind. In *Proceedings of the Fifth European African Conference on Wind Engineering, Florence, Italy, July 19th–23rd*, 2009.

[54] L. M. Cléon and A. Jourdain. Protection of line ln5 against cross winds. In *In Proceedings of the 5th World Congress on Railway Research, Köln, Germany*, 2001.

[55] S. A. Coleman and C. J. Baker. High sided road vehicles in cross winds. *Journal of Wind Engineering and Industrial Aerodynamics*, 36:1383–1392, 1990.

[56] S. A. Coleman and C. J. Baker. An experimental study of the aerodynamic behaviour of high sided lorries in cross winds. *Journal of Wind Engineering and Industrial Aerodynamics*, 53(3):401–429, 1994.

[57] S. Coles, J. Bawa, L. Trenner, and P. Dorazio. *An introduction to statistical modeling of extreme values*, volume 208. Springer, 2001.

[58] R. K. Cooper. Probability of trains overturning in high winds. In *Wind Engineering, Proceedings of the International Conference, 5th, Colorado State University, Fort Collins, July 8-14, 1979.*, volume 2, 1980.

[59] R. K. Cooper. Atmospheric turbulence with respect to moving ground vehicles. *Journal of wind engineering and industrial aerodynamics*, 17(2):215–238, 1984.

[60] J. M. Copley. The three-dimensional flow around railway trains. *Journal of Wind Engineering and Industrial Aerodynamics*, 26(1):21–52, 1987.

[61] Q. Dai and R. K. Young. High wind warning system to prevent overturning truck crashes in wyoming. In *Transportation Research Board 2010 Annual Meeting, Washington, D.C., USA*, 2010.

[62] I. Daubechies. *Ten lectures on wavelets*, volume 61. Society for Industrial and Applied Mathematics (SIAM), 1992.

[63] A. G. Davenport. The application of statistical concepts to the wind loading of structures. In *ICE Proceedings*, volume 19, pages 449–472. Thomas Telford, 1961.

[64] A. C. Davison and R. L. Smith. Models for exceedances over high thresholds. *Journal of the Royal Statistical Society. Series B (Methodological)*, pages 393–442, 1990.

[65] D. Delaunay, C. Baker, F. Cheli, H. Morvan, L. Berger, M. Casazza, C. Gomez, C. Cleac'h, R. Saffell, R. Grégoire, et al. Development of wind alarm systems for road and rail vehicles: presentation of the weather project. *Proceedings of the SIRWEC*, 2006.

[66] D. Delaunay and J. P. Locatelly. A gust model for the design of large horizontal axis wind turbines: completion and validation. *In: Proc. European Community Wind Energy Conference, Madrid, Spain*, pages 176–180, 1990.

[67] G. Deodatis and M. Shinozuka. Auto-regressive model for nonstationary stochastic processes. *Journal of engineering mechanics*, 114(11):1995–2012, 1988.

[68] A. Der Kiureghian. The geometry of random vibrations and solutions by form and sorm. *Probabilistic Engineering Mechanics*, 15(1):81–90, 2000.

[69] DeuFraKo-Projekt. Wind model developed in deufrako x-wind project. In *DeuFraKo TSI meeting, Brussels*, 2003.

[70] AG Deutsche Bahn. Handbuch für den sicherheitsnachweis bei seitenwind. *Deutsche Bahn AG, DB Sytemtechnik, TZF*, 102, 2001.

[71] M. Di Paola. Digital simulation of wind field velocity. *Journal of Wind Engineering and Industrial Aerodynamics*, 74:91–109, 1998.

[72] G. Diana, M. Burlando, F. Cheli, A. Freda, C. F. Ratto, D. Rocchi, G. Solari, M. Testa, and G. Tomasini. A new methodology to perform the risk analysis of cross wind on high speed lines. *Proceedings of the 8th WCRR, May*, pages 18–22, 2008.

[73] B. Diedrichs. *Computational methods for crosswind stability of railway trains: A literature survey.* Department of Aeronautical and Vehicle Engineering, Royal Institute of Technology, 2005.

[74] Y. Ding, M. Sterlings, and C. Baker. An alternative approach to modelling train stability in high cross winds. *Proceedings of the Institution of Mechanical Engineers, Part F: Journal of Rail and Rapid Transit*, 222(1):85–97, 2008.

[75] E. Donges. A two-level model of driver steering behavior. *Human Factors: The Journal of the Human Factors and Ergonomics Society*, 20(6):691–707, 1978.

[76] A. Dorvlo. Estimating wind speed distribution. *Energy Conversion and Management*, 43(17):2311–2318, 2002.

[77] H. Dugoff, P. Fancher, and L. Segel. An analysis of tire traction properties and their influence on vehicle dynamic performance. *SAE paper*, 700377:1219–1243, 1970.

[78] S. El Adlouni, T. Ouarda, X. Zhang, R. Roy, and B. Bobée. Generalized maximum likelihood estimators for the nonstationary generalized extreme value model. *Water Resources Research*, 43(3), 2007.

[79] P. Embrechts, C. Klüppelberg, and T. Mikosch. *Modelling extremal events: for insurance and finance*, volume 33. Springer, 1997.

[80] C. Esveld. Modern railway track. *NS Permanent Way Department*, 1989.

[81] Technical Committee CEN/TC 250 "Structural Eurocodes". Eurocode 1: Actions on structure - part 1-1: General actions - densities, self-weight, imposed loads for buildings. *Norm European Standard*, EN 1991-1-1.

[82] J. Franke, W. K. Härdle, and C. M. Hafner. *Statistics of Financial Markets: An Introduction.* Springer, 2011.

[83] T. T. Fujita. Downbursts: meteorological features and wind field characteristics. *Journal of Wind Engineering and Industrial Aerodynamics*, 36:75–86, 1990.

[84] PE Gautier, Th Tielkes, F Sourget, E Allain, M Grab, and Ch Heine. Strong wind risks in railways: the deufrako crosswind program. In *Proc. of WCRR*, pages 463–475, 2003.

[85] R. Glaus. *The Swiss Trolley: A Modular System for Track Surveying.* Institut für Geodäsie und Photogrammetrie, 2006.

[86] G. H. Goedecke, V. E. Ostashev, D. K. Wilson, and H. J. Auvermann. Quasi-wavelet model of von kármán spectrum of turbulent velocity fluctuations. *Boundary-layer meteorology*, 112(1):33–56, 2004.

[87] R. W. Goldman, M. El-Gindy, and B. T. Kulakowski. Rollover dynamics of road vehicles: Literature survey. *International Journal of Heavy Vehicle Systems*, 8(2):103–141, 2001.

[88] M. Grigoriu and E. Harper. *Applied non-Gaussian processes: Examples, theory, simulation, linear random vibration, and MATLAB solutions.* PTR Prentice Hall Englewood Cliffs, 1995.

[89] S. D. Grimshaw. Computing maximum likelihood estimates for the generalized pareto distribution. *Technometrics*, 35(2):185–191, 1993.

[90] W. H. Guo and Y. L. Xu. Safety analysis of moving road vehicles on a long bridge under crosswind. *Journal of Engineering Mechanics*, 132(4):438–446, 2006.

[91] K. Gurley and A. Kareem. Applications of wavelet transforms in earthquake, wind and ocean engineering. *Engineering structures*, 21(2):149–167, 1999.

[92] K. R. Gurley, M. A. Tognarelli, and A. Kareem. Analysis and simulation tools for wind engineering. *Probabilistic Engineering Mechanics*, 12(1):9–31, 1997.

[93] A. Hac, D. Fulk, and H. Chen. Stability and control considerations of vehicle-trailer combination. *SAE SP*, 2188:125, 2008.

[94] J. D. Hamilton. *Time series analysis*, volume 2. Princeton university press Princeton, 1994.

[95] X. He, J. Fang, A. Scanlon, and Z. Chen. Wavelet-based nonstationary wind speed model in dongting lake cable-stayed bridge. *Engineering*, 2(11):895–903, 2010.

[96] H. N. Hemida. *Large-eddy simulation of the flow around simplified high-speed trains under side wind conditions.* PhD thesis, Chalmers University of Technology, Goteborg, Sweden, 2006.

[97] R. A. Hess and A. Modjtahedzadeh. A control theoretic model of driver steering behavior. *Control Systems Magazine, IEEE*, 10(5):3–8, 1990.

[98] G. J. Holland. An analytic model of the wind and pressure profiles in hurricanes. *Monthly weather review*, 108(8):1212–1218, 1980.

[99] J. D. Holmes. Non-gaussian characteristics of wind pressure fluctuations. *Journal of Wind Engineering and Industrial Aerodynamics*, 7(1):103–108, 1981.

[100] J.D. Holmes and W.W. Moriarty. Application of the generalized pareto distribution to extreme value analysis in wind engineering. *Journal of Wind Engineering and Industrial Aerodynamics*, 83(1):1–10, 1999.

[101] N. E. Huang, X. Chen, M. T. Lo, and Z. Wu. On hilbert spectral representation: a true time-frequency representation for nonlinear and nonstationary data. *Advances in Adaptive Data Analysis*, 3(01n02):63–93, 2011.

[102] N. E. Huang and S. S. Shen. *Hilbert-Huang transform and its applications*, volume 5. World Scientific, 2005.

[103] N. E. Huang, Z. Shen, S. R. Long, M. C. Wu, H. H. Shih, Q. Zheng, N. C. Yen, C. C. Tung, and H. H. Liu. The empirical mode decomposition and the hilbert spectrum for nonlinear and non-stationary time series analysis. *Proc. R. Soc. Lond.*, 454A:903–995, 1998.

[104] N. E. Huang, M. Wu, W. Qu, S. R. Long, and S. Shen. Applications of hilbert–huang transform to non-stationary financial time series analysis. *Applied stochastic models in business and industry*, 19(3):245–268, 2003.

[105] N. E. Huang, Z. Wu, S. R. Long, K. C. Arnold, X. Chen, and K. Blank. On instantaneous frequency. *Advances in Adaptive Data Analysis*, 1(02):177–229, 2009.

[106] M. Hubner, T. Stork, U. Becker, and E. Schnieder. Lateral stabilization of vehicle-trailer combinations against crosswind disturbances by means of sliding control. In *Control and Automation, 2008 16th Mediterranean Conference on*, pages 431–438. IEEE, 2008.

[107] K. Hudha, F. Ahmad, Z. A. Kadir, and H. Jamaluddin. Pid controller with roll moment rejection for pneumatically actuated active roll control (arc) suspension system.

[108] Y. Hundecha, A. St-Hilaire, T. Ouarda, S. El Adlouni, and P. Gachon. A nonstationary extreme value analysis for the assessment of changes in extreme annual wind speed over the gulf of st. lawrence, canada. *Journal of Applied Meteorology and Climatology*, 47(11):2745–2759, 2008.

[109] Z. M. Hussain, A. Z. Sadik, and P. O'Shea. *Digital signal processing: an introduction with MATLAB and applications*. Springer, 2011.

[110] M. Islam, X. Ding, and Y. He. A closed-loop dynamic simulation-based design method for articulated heavy vehicles with active trailer steering systems. *Vehicle System Dynamics*, 50(5):675–697, 2012.

[111] P. Jain. *Wind energy engineering.* McGraw-Hill, New York, 2011.

[112] L. Jiang, W. Yuan, and Z. Yang. Stationary wavelet-based analysis and simulation of unsteady wind in aeolian sand transport. In *9th International Conference on Electronic Measurement & Instruments*, pages 789–794. IEEE, 2009.

[113] Magnus Juhlin. *Assessment of crosswind performance of buses.* PhD thesis, KTH, 2009.

[114] C. Justus, W. Hargraves, A. Mikhail, and D. Graber. Methods for estimating wind speed frequency distributions. *Journal of Applied Meteorology*, 17(3):350–353, 1978.

[115] J. Kalker. A fast algorithm for the simplified theory of rolling contact. *Vehicle System Dynamics*, 11(1):1–13, 1982.

[116] A. Kareem. Numerical simulation of wind effects: a probabilistic perspective. *Journal of Wind Engineering and Industrial Aerodynamics*, 96(10):1472–1497, 2008.

[117] A. Kareem and T. Wu. Wind-induced effects on bluff bodies in turbulent flows: Nonstationary, non-gaussian and nonlinear features. *Journal of Wind Engineering and Industrial Aerodynamics*, 122:21–37, 2013.

[118] A. Kareem and J. Zhao. Analysis of non-gaussian surge response of tension leg platforms under wind loads. *Journal of Offshore Mechanics and Arctic Engineering*, 116(3):137–144, 1994.

[119] H. Kiefer. *Windlasten an quaderförmigen Gebäuden in bebauten Gebieten.* PhD thesis, Karlsruhe, Univ., Diss., 2003.

[120] W. Kijawatworawet, H. J. Pradlwarter, and G. I. Schuëller. Structural reliability estimation by adaptive importance directional sampling. In *Proceedings of the 7th International Conference on Structural Safety and Reliability (ICOSSAR'97)*, pages 891–897, 1998.

[121] H. Kim and P. Durbin. Observations of the frequencies in a sphere wake and of drag increase by acoustic excitation. *Physics of Fluids (1958-1988)*, 31(11):3260–3265, 1988.

[122] T. Kitagawa and T. Nomura. A wavelet-based method to generate artificial wind fluctuation data. *Journal of wind engineering and industrial aerodynamics*, 91(7):943–964, 2003.

[123] P. Koutsourelakis, H. Pradlwarter, and G. Schuëller. Reliability of structures in high dimensions, part 1: algorithms and applications. *Probabilistic Engineering Mechanics*, 19(4):409–417, 2004.

[124] L. Kristensen, M. Casanova, M. S. Courtney, and I. Troen. In search of a gust definition. *Boundary-Layer Meteorology*, 55(1-2):91–107, 1991.

[125] M. Kumar and C. K. Strong. *Comparative evaluation of automated wind warning systems.* Western Transportation Institute, 2006.

[126] H. Kwon, Y. Park, D. Lee, and M. Kim. Wind tunnel experiments on korean high-speed trains using various ground simulation techniques. *Journal of Wind Engineering and Industrial Aerodynamics*, 89(13):1179–1195, 2001.

[127] G. C. Larsen, W. Bierbooms, and K. S. Hansen. Mean gust shapes. Technical report, 2003.

[128] G. C. Larsen and K. S. Hansen. *Database on wind characteristics-analyses of wind turbine design loads.* Pitney Bowes Management Services Denmar, 2004.

[129] F. B. Leahy. Discrete gust model for launch vehicle assessments. In *12th Conference on Aviation, Range and Aerospace Metrology*, 2006.

[130] D. Lee and A. Yamamoto. Wavelet analysis: Theory and applications. *Hewlett Packard journal*, 45:44–52, 1994.

[131] J. Lee Rodgers and W. Nicewander. Thirteen ways to look at the correlation coefficient. *The American Statistician*, 42(1):59–66, 1988.

[132] C. Li, J. Li, and J. Shen. Simulation of transient non-stationary winds. In *Fifth International Joint Conference on INC, IMS and IDC*, pages 1509–1512. IEEE, 2009.

[133] X. Li, editor. *Green energy : basic concepts and fundamentals.* Progress in green energy. Springer, London, 2011.

[134] M. Lindahl. Track geometry for high-speed railways. Technical Report TRITA–FKT 2001:54, Royal Institute of Technology–Railway Technology, 2001.

[135] P. Liu and A. Der Kiureghian. Multivariate distribution models with prescribed marginals and covariances. *Probabilistic Engineering Mechanics*, 1(2):105–112, 1986.

[136] P. Liu and A. Der Kiureghian. Optimization algorithms for structural reliability. *Structural safety*, 9(3):161–177, 1991.

[137] J. Maeda and M. Makino. Characteristics of gusty winds simulated by an arma model. *Journal of Wind Engineering and Industrial Aerodynamics*, 41(1):427–436, 1992.

[138] G. Mantriota. A theoretical study on the stability of car-caravan systems under the influence of crosswind. *International Journal of Vehicle Systems Modelling and Testing*, 1(1):168–191, 2005.

[139] J. F. Manwell, J. G. McGowan, and A. L. Rogers. *Wind energy explained : theory, design and application.* Wiley, Chichester, repr. with corr. edition, 2005. Includes bibliographical references and index.

[140] R. Marino, S. Scalzi, G. Orlando, and M. Netto. A nested pid steering control for lane keeping in vision based autonomous vehicles. In *American Control Conference*, pages 2885–2890. IEEE, 2009.

[141] B. Mashadi, M. Mahmoudi-Kaleybar, P. Ahmadizadeh, and A. Oveisi. A path-following driver/vehicle model with optimized lateral dynamic controller. *Latin American Journal of Solids and Structures*, 11(4):613–630, 2014.

[142] G. Matschke and C. Heine. Full scale tests on side wind effects on trains. evaluation of aerodynamic coefficients and efficiency of wind breaking devices. *TRANSAERO*, 79:27–38, 2002.

[143] J. McCarthy. *Evaluation of intelligent transportation system alternatives for reducing the risks of truck rollover crashes due to high winds.* Technical report, U.S. Department of Transportation, Federal Highway Administration, FHWA-WY-07/01F, 2007.

[144] C. Moon and S. Choi. A driver model for vehicle lateral dynamics. *International Journal of Vehicle Design*, 56(1):49–80, 2011.

[145] E. Morfiadakis, G. Glinou, and M. Koulouvari. The suitability of the von karman spectrum for the structure of turbulence in a complex terrain wind farm. *Journal of wind engineering and industrial aerodynamics*, 62(2):237–257, 1996.

[146] M. D. Morris. Factorial sampling plans for preliminary computational experiments. *Technometrics*, 33(2):161–174, 1991.

[147] G. P. Nason and T. Sapatinas. Wavelet packet transfer function modelling of non-stationary time series. *Statistics and Computing*, 12(1):45–56, 2002.

[148] A. Nataf. Détermination des distributions de probabilités dont les marges sont données. *Comptes rendus de l'academie des sciences*, 225(1):42–43, 1962.

[149] N. Nguyen Minh, T. Miyata, H. Yamada, and Y. Sanada. Numerical simulation of wind turbulence and buffeting analysis of long-span bridges. *Journal of Wind Engineering and Industrial Aerodynamics*, 83(1):301–315, 1999.

[150] J. Nie and B. R. Ellingwood. Directional methods for structural reliability analysis. *Structural Safety*, 22(3):233–249, 2000.

[151] A. O'Dwyer. *Handbook of PI and PID controller tuning rules*, volume 2. World Scientific, 2006.

[152] H. B. Pacejka and E. Bakker. The magic formula tyre model. *Vehicle system dynamics*, 21(S1):1–18, 1992.

[153] J. P. Palutikof, B. B. Brabson, D. H. Lister, and S. T. Adcock. A review of methods to calculate extreme wind speeds. *Meteorological applications*, 6(02):119–132, 1999.

[154] Z. Peng, P. W. Tse, and F. Chu. An improved hilbert–huang transform and its application in vibration signal analysis. *Journal of Sound and Vibration*, 286(1):187–205, 2005.

[155] J. A. Peterka and J. E. Cermak. Wind pressures on buildings-probability densities. *Journal of the structural division*, 101(6):1255–1267, 1975.

[156] M. Plöchl and J. Edelmann. Driver models in automobile dynamics application. *Vehicle System Dynamics*, 45(7-8):699–741, 2007.

[157] A. Popoola, S. Ahmad, and K. Ahmad. A fuzzy-wavelet method for analyzing non-stationary time series. In *Proc. of The 5th Int. Conf. on Recent Advances in Soft Computing*. Citeseer, 2004.

[158] M. B. Priestley. Evolutionary spectra and non-stationary processes. *Journal of the Royal Statistical Society. Series B (Methodological)*, pages 204–237, 1965.

[159] G. Prokop. Modeling human vehicle driving by model predictive online optimization. *Vehicle System Dynamics*, 35(1):19–53, 2001.

[160] C. Proppe and C. Wetzel. Overturning probability of railway vehicles under wind gust loads. In *Iutam Symposium on Dynamics and Control of Nonlinear Systems with Uncertainty*, pages 23–32. Springer, 2007.

[161] C. Proppe and C. Wetzel. A probabilistic approach for assessing the crosswind stability of ground vehicles. *Vehicle System Dynamics*, 48(S1):411–428, 2010.

[162] C. Proppe and X. Zhang. Risk analysis of railway vehicles under strong crosswinds. *The international journal of railway Technology*, 2:93–111, 2013.

[163] E. Offei R. Young and Q. Dai. High wind warning system for bordeaux, wyoming. Technical report, Department of Civil and Architectural Engineering, University of Wyoming, 2010.

[164] R. Rackwitz. Reliability analysis–a review and some perspectives. *Structural safety*, 23(4):365–395, 2001.

[165] N. Rahman and M. Alam. Active vibration control of a piezoelectric beam using pid controller: Experimental study. *Latin American Journal of Solids and Structures*, 9(6):657–673, 2012.

[166] K. Rangavajhula and H. Tsao. Command steering of trailers and command-steering-based optimal control of an articulated system for tractor-track following. *Proceedings of the Institution of Mechanical Engineers, Part D: Journal of Automobile Engineering*, 222(6):935–954, 2008.

[167] R. D. Reiss and M. Thomas. *Statistical analysis of extreme values: with applications to insurance, finance, hydrology and other fields*. Springer, 2007.

[168] S. O. Rice. Mathematical analysis of random noise. *Bell System Technical Journal*, 24(1):46–156, 1945.

[169] M. Rosenblatt. Remarks on a multivariate transformation. *The annals of mathematical statistics*, pages 470–472, 1952.

[170] R. Rossi, M. Lazzari, and R. Vitaliani. Wind field simulation for structural engineering purposes. *International journal for numerical methods in engineering*, 61(5):738–763, 2004.

[171] A. G. Ryan. *The simulation of transient cross-wind gusts and their aerodynamic influence on passenger cars.* PhD thesis, Durham E-Theses, Durham University, 2000.

[172] H. Sakamoto and H. Haniu. A study on vortex shedding from spheres in a uniform flow. *Journal of Fluids Engineering*, 112(4):386–392, 1990.

[173] A. Saltelli, M. Ratto, T. Andres, F. Campolongo, J. Cariboni, D. Gatelli, M. Saisana, and S. Tarantola. *Global sensitivity analysis: the primer.* Wiley. com, 2008.

[174] S. Sanquer, C. Barre, M. D. de Virel, and L. M. Cleon. Effect of cross winds on high-speed trains: development of a new experimental methodology. *Journal of Wind Engineering and Industrial Aerodynamics*, 92(7):535–545, 2004.

[175] A. Sarkar, S. Singh, and D. Mitra. Wind climate modeling using weibull and extreme value distribution. *International Journal of Engineering, Science and Technology*, 3(5), 2011.

[176] G. Schuëller, H. Pradlwarter, and P. Koutsourelakis. A critical appraisal of reliability estimation procedures for high dimensions. *Probabilistic Engineering Mechanics*, 19(4):463–474, 2004.

[177] J. Seguro and T. Lambert. Modern estimation of the parameters of the weibull wind speed distribution for wind energy analysis. *Journal of Wind Engineering and Industrial Aerodynamics*, 85(1):75–84, 2000.

[178] J. D. Setiawan, M. Safarudin, and A. Singh. Modeling, simulation and validation of 14 dof full vehicle model. In *Instrumentation, Communications, Information Technology, and Biomedical Engineering (ICICI-BME), 2009 International Conference on*, pages 1–6. IEEE, 2009.

[179] H. Shariatmadar and Golsa. Behnamrad. Active control of structures and reliability analysis by subset simulation method. *Nationalpark-Forschung in der Schweiz-Switzerland Research Park Journal*, 102, 2013.

[180] T. Shim and C. Ghike. Understanding the limitations of different vehicle models for roll dynamics studies. *Vehicle system dynamics*, 45(3):191–216, 2007.

[181] M. Shimamura and N. Kobayashi. Development of a strong wind warning system. *Japanese railway engineering*, 43(1), 2003.

[182] M. Shinozuka and G. Deodatis. Simulation of stochastic processes by spectral representation. *Applied Mechanics Reviews*, 44:191, 1991.

[183] S. Si, K. Xu, and R. Li. *Mathematical modeling (in Chinese)*. Naval Aeronautical Institute of Technology, 2003.

[184] E. Simiu and R. H. Scanlan. Wind effects on structures: An introduction to wind engineering, 1986.

[185] N. D. Smith. Understanding parameters influencing tire modeling. *Colorado State University*, 2004.

[186] J. Snæbjörnsson, C. Baker, and R. Sigbjörnsson. Probabilistic assessment of road vehicle safety in windy environments. *Journal of Wind Engineering and Industrial Aerodynamics*, 95(9):1445–1462, 2007.

[187] S. Song, Z. Lu, and H. Qiao. Subset simulation for structural reliability sensitivity analysis. *Reliability Engineering & System Safety*, 94(2):658–665, 2009.

[188] M. Sterling, C. Baker, A. Bouferrouk, H. ONeil, S. Wood, and E. Crosbie. An investigation of the aerodynamic admittances and aerodynamic weighting functions of trains. *Journal of Wind Engineering and Industrial Aerodynamics*, 97(11):512–522, 2009.

[189] M. Sterling, A. D. Quinn, D. M. Hargreaves, F. Cheli, E. Sabbioni, G. Tomasini, D. Delaunay, C. J. Baker, and H. Morvan. A comparison of different methods to evaluate the wind induced forces on a high sided lorry. *Journal of Wind Engineering and Industrial Aerodynamics*, 98(1):10–20, 2010.

[190] M. Suzuki, K. Tanemoto, and T. Maeda. Aerodynamic characteristics of train/vehicles under cross winds. *Journal of Wind Engineering and Industrial Aerodynamics*, 91(1):209–218, 2003.

[191] J. Svendenius. *Tire modeling and friction estimation*. PhD thesis, Lund University, 2007.

[192] S. H. Tamaddoni and S. Taheri. Yaw stability control of tractor semi-trailers. *SAE (08CV-0019)*, 2008.

[193] P. Teufel. Böenmodellierung und lastabminderung für ein flexibles flugzeug. 2003.

[194] Associated Press theguardian.com. Lorries were overturned in toyama city. Online, Version: 17.04.2014. http://www.theguardian.com/world/2012/apr/04/japan-storm-honshu-tokyo-nuclear.

[195] D. Thomas. Lateral stability of high-speed trains at unsteady crosswind. 2009.

[196] D. Thomas. *On Rail Vehicle Dynamics in Unsteady Crosswind Conditions.* PhD thesis, KTH, 2013.

[197] J. L. Torres, A. Garcia, M. De Blas, and A. De Francisco. Forecast of hourly average wind speed with arma models in navarre (spain). *Solar Energy*, 79(1):65–77, 2005.

[198] P. W. Tse, W. Yang, and H. Y. Tam. Machine fault diagnosis through an effective exact wavelet analysis. *Journal of Sound and Vibration*, 277(4):1005–1024, 2004.

[199] L. Wang and A. Kareem. Modeling of non-stationary winds in gust-fronts. In *9th ASCE Specialty Conference on Probabilistic Mechanics and Structural Reliability*, 2004.

[200] Z. Wang. Simulation of wind loading. *Journal of Building Structures*, 15(1):44–52, 1994.

[201] William W. S. Wei. *Time series analysis.* Addison-Wesley publ, 1994.

[202] C. Wetzel. *Zur probabilistischen Betrachtung von Schienen-und Kraftfahrzeugsystemen unter zufälliger Windanregung*, volume 12. KIT Scientific Publishing, 2009.

[203] C. Wetzel and C. Proppe. Crosswind stability of high-speed trains: A stochastic approach. In *July, Milano, Italy. BBAA VI International Colloquium on: Bluff Bodies Aerodynamics & Applications*, 2008.

[204] C. Wetzel and C. Proppe. On reliability and sensitivity methods for vehicle systems under stochastic crosswind loads. *Vehicle System Dynamics*, 48(1):79–95, 2010.

[205] C. Wetzel and C. Proppe. Stochastic modeling in multibody dynamics: aerodynamic loads on ground vehicles. *Journal of computational and nonlinear dynamics*, 5(3), 2010.

[206] A. H. Wickens. Fundamentals of rail vehicle dynamics, 2003.

[207] J. Wiernga. Representative roughness parameters for homogeneous terrain. *Boundary-Layer Meteorology*, 63(4):323–363, 1993.

[208] D. Wu. A theoretical study of the yaw/roll motions of a multiple steering articulated vehicle. *Proceedings of the Institution of Mechanical Engineers, Part D: Journal of Automobile Engineering*, 215(12):1257–1265, 2001.

[209] Z. Wu and N. E. Huang. Ensemble empirical mode decomposition: a noise-assisted data analysis method. *Advances in adaptive data analysis*, 1(01):1–41, 2009.

[210] H. Xu and Y. Gao. *Fundamentals of automobile application engineering (in Chinese).* Tsinghua University Press, 2004.

[211] Y. Xu. *Wind Effects on Cable-supported Bridges.* John Wiley & Sons, 2013.

[212] Y. Xu and J. Chen. Characterizing nonstationary wind speed using empirical mode decomposition. *Journal of structural Engineering*, 130(6):912–920, 2004.

[213] M. Yamada and K. Ohkitani. Orthonormal wavelet analysis of turbulence. *Fluid Dynamics Research*, 8(1):101–115, 1991.

[214] D. H. Yeo. Numerical simulation of along-wind loading on small structures using a simplified wind flow model. *NIST Technical Note 1683*, 2010.

[215] R. K. Young. Use of real time weather variables to operate rural interstates in wyoming. In *SIRWEC 2012, Helsinki, 23-25 May*, 2012.

[216] L. Zhang and T. Zhang. General nonstationary random input model of road surface wity four wheels correlated. *Journal of Vibration and Shock*, 7:019, 2008.

[217] X. Zhang, K. Lai, and S. Wang. A new approach for crude oil price analysis based on empirical mode decomposition. *Energy Economics*, 30(3):905–918, 2008.

[218] J. Zhong. Pid controller tuning: A short tutorial. *Class lesson, Purdue University*, 2006.

KARLSRUHER INSTITUT FÜR TECHNOLOGIE (KIT)
SCHRIFTENREIHE DES INSTITUTS FÜR TECHNISCHE MECHANIK (ITM)

ISSN 1614-3914

Die Bände sind unter www.ksp.kit.edu als PDF frei verfügbar oder als Druckausgabe zu bestellen.

Band 1 **Marcus Simon**
Zur Stabilität dynamischer Systeme mit stochastischer Anregung. 2004
ISBN 3-937300-13-9

Band 2 **Clemens Reitze**
Closed Loop, Entwicklungsplattform für mechatronische Fahrdynamikregelsysteme. 2004
ISBN 3-937300-19-8

Band 3 **Martin Georg Cichon**
Zum Einfluß stochastischer Anregungen auf mechanische Systeme. 2006
ISBN 3-86644-003-0

Band 4 **Rainer Keppler**
Zur Modellierung und Simulation von Mehrkörpersystemen unter Berücksichtigung von Greifkontakt bei Robotern. 2007
ISBN 978-3-86644-092-0

Band 5 **Bernd Waltersberger**
Strukturdynamik mit ein- und zweiseitigen Bindungen aufgrund reibungsbehafteter Kontakte. 2007
ISBN 978-3-86644-153-8

Band 6 **Rüdiger Benz**
Fahrzeugsimulation zur Zuverlässigkeitsabsicherung von karosseriefesten Kfz-Komponenten. 2008
ISBN 978-3-86644-197-2

Band 7 **Pierre Barthels**
Zur Modellierung, dynamischen Simulation und Schwingungsunterdrückung bei nichtglatten, zeitvarianten Balkensystemen. 2008
ISBN 978-3-86644-217-7

KARLSRUHER INSTITUT FÜR TECHNOLOGIE (KIT)
SCHRIFTENREIHE DES INSTITUTS FÜR TECHNISCHE MECHANIK (ITM)

ISSN 1614-3914

Band 8 **Hartmut Hetzler**
Zur Stabilität von Systemen bewegter Kontinua mit Reibkontakten am Beispiel des Bremsenquietschens. 2008
ISBN 978-3-86644-229-0

Band 9 **Frank Dienerowitz**
Der Helixaktor – Zum Konzept eines vorverwundenen Biegeaktors. 2008
ISBN 978-3-86644-232-0

Band 10 **Christian Rudolf**
Piezoelektrische Self-sensing-Aktoren zur Korrektur statischer Verlagerungen. 2008
ISBN 978-3-86644-267-2

Band 11 **Günther Stelzner**
Zur Modellierung und Simulation biomechanischer Mehrkörpersysteme. 2009
ISBN 978-3-86644-340-2

Band 12 **Christian Wetzel**
Zur probabilistischen Betrachtung von Schienen- und Kraftfahrzeugsystemen unter zufälliger Windanregung. 2010
ISBN 978-3-86644-444-7

Band 13 **Wolfgang Stamm**
Modellierung und Simulation von Mehrkörpersystemen mit flächigen Reibkontakten. 2011
ISBN 978-3-86644-605-2

Band 14 **Felix Fritz**
Modellierung von Wälzlagern als generische Maschinenelemente einer Mehrkörpersimulation. 2011
ISBN 978-3-86644-667-0

KARLSRUHER INSTITUT FÜR TECHNOLOGIE (KIT)
SCHRIFTENREIHE DES INSTITUTS FÜR TECHNISCHE MECHANIK (ITM)

ISSN 1614-3914

Band 15 **Aydin Boyaci**
Zum Stabilitäts- und Bifurkationsverhalten hochtouriger Rotoren in Gleitlagern. 2012
ISBN 978-3-86644-780-6

Band 16 **Rugerri Toni Liong**
Application of the cohesive zone model to the analysis of rotors with a transverse crack. 2012
ISBN 978-3-86644-791-2

Band 17 **Ulrich Bittner**
Strukturakustische Optimierung von Axialkolbeneinheiten. Modellbildung, Validierung und Topologieoptimierung. 2013
ISBN 978-3-86644-938-1

Band 18 **Alexander Karmazin**
Time-efficient Simulation of Surface-excited Guided Lamb Wave Propagation in Composites. 2013
ISBN 978-3-86644-935-0

Band 19 **Heike Vogt**
Zum Einfluss von Fahrzeug- und Straßenparametern auf die Ausbildung von Straßenunebenheiten. 2013
ISBN 978-3-7315-0023-0

Band 20 **Laurent Ineichen**
Konzeptvergleich zur Bekämpfung der Torsionsschwingungen im Antriebsstrang eines Kraftfahrzeugs. 2013
ISBN 978-3-7315-0030-8

Band 21 **Sietze van Buuren**
Modeling and simulation of porous journal bearings in multibody systems. 2013
ISBN 978-3-7315-0084-1

KARLSRUHER INSTITUT FÜR TECHNOLOGIE (KIT)
SCHRIFTENREIHE DES INSTITUTS FÜR TECHNISCHE MECHANIK (ITM)

ISSN 1614-3914

Band 22 **Dominik Kern**
Neuartige Drehgelenke für reibungsarme Mechanismen. 2013
ISBN 978-3-7315-0103-9

Band 23 **Nicole Gaus**
Zur Ermittlung eines stochastischen Reibwerts und dessen Einfluss auf reibungserregte Schwingungen. 2013
ISBN 978-3-7315-0118-3

Band 24 **Fabian Bauer**
Optimierung der Energieeffizienz zweibeiniger Roboter durch elastische Kopplungen. 2014
ISBN 978-3-7315-0256-2

Band 25 **Benedikt Wiegert**
Nichtlineare Schwingungen von Systemen mit elastohydrodynamischen Linienkontakten. 2015
ISBN 978-3-7315-0350-7

Band 26 **Arsenty Tikhomolov**
Analytische, numerische und messtechnische Untersuchung der Dynamik von Fahrzeugkupplungen am Beispiel des Trennproblems. 2015
ISBN 978-3-7315-0362-0

Band 27 **Daniel Maier**
On the Use of Model Order Reduction Techniques for the Elastohydrodynamic Contact Problem. 2015
ISBN 978-3-7315-0369-9

Band 28 **Xiaoyu Zhang**
Crosswind stability of vehicles under nonstationary wind excitation. 2015
ISBN 978-3-7315-0376-7